Rapid Qualitative Inquiry

Rapid Qualitative Inquiry

A Field Guide to Team-Based Assessment

Second Edition

James Beebe

Rowman & Littlefield
Lanham • Boulder • New York • London

Published by Rowman & Littlefield
A wholly owned subsidiary of The Rowman & Littlefield Publishing Group, Inc.
4501 Forbes Boulevard, Suite 200, Lanham, Maryland 20706
www.rowman.com

16 Carlisle Street, London W1D 3BT, United Kingdom

First edition 2001

British Library Cataloguing in Publication Information Available

Library of Congress Cataloging-in-Publication Data
Beebe, James.
[Rapid assessment process]
Rapid qualitative inquiry : a field guide to team-based assessment / James Beebe. — Second edition.
pages cm
Revised edition of the author's Rapid assessment process : an introduction, published in 2001.
Includes bibliographical references and index.
ISBN 978-0-7591-2319-9 (cloth : alk. paper) — ISBN 978-0-7591-2320-5 (pbk. : alk. paper) — ISBN 978-0-7591-2321-2 (electronic) 1. Evaluation research (Social action programs) 2. Social sciences—Methodology. 3. Qualitative research. I. Title.
H62.B3537 2014
001.4'2—dc23

2014027856

∞™ The paper used in this publication meets the minimum requirements of American National Standard for Information Sciences—Permanence of Paper for Printed Library Materials, ANSI/NISO Z39.48-1992.

Printed in the United States of America

In memory of Robert B. Textor,
1923 to 2013
Advisor, Advocate, Colleague

BRIEF CONTENTS

DETAILED CONTENTS

TABLES AND FIGURES

Tables

Figures

PREFACE TO THE SECOND EDITION

The Journey That Resulted in This Book

When I was a Peace Corps volunteer in the Philippines I quickly learned that I needed to know the meanings people there attached to words, because I could not assume we assigned the same meanings to the same words. Even something as straightforward as offering a soft drink to someone was subject to misunderstanding and embarrassment, regardless of whether the offer was made in the local language or in English. The offer was made, the response was "Thank you," and my expectation was that the offer had been accepted. I delivered the soft drink to someone who did not want it and discovered that the "Thank you" was gratitude for the offer and not an acceptance of the offer. Graduate studies at the University of the Philippines and Stanford, including a year of fieldwork in a village in the Philippines, introduced me to the vocabulary and the methodology of anthropology and helped me begin to understand the requirements for more successful cross-cultural communication. My approach to qualitative research was significantly influenced by Professor E. Arsenio Manuel (1909–2003), Anthropology Department, University of the Philippines, and Professor Robert B. Textor (1923–2013) Anthropology Department and the School of Education, Stanford University.

After graduate school, I joined the U.S. Agency for International Development (USAID) and spent the next eighteen years as a practitioner helping to implement foreign assistance projects. I spent more than fourteen of these years overseas, with long-term assignments in Sudan, the Philippines, Liberia, and South Africa. These experiences reinforced for me the need to pay attention to the categories used by others and the meanings they attached to the words they used (see ch. 2, "Emic and Etic"). I soon realized that successful interventions have to be based on genuine partnership between outsiders who often control access to resources and the insiders who will ultimately be responsible for implementing changes. Partnership in designing interventions is critical for their success. The potential for miscommunication is always present when outsiders and insiders attempt to collaborate on the design of an intervention as well as on implementation, monitoring, and evaluation. To minimize miscommunication everyone needs to recognize that categories and meaning are socially constructed. Meanings of words depend upon their cultural context. We ignore the context at our peril.

Qualitative research can improve the communication process, but often neither the time nor other resources are available for traditional approaches to this type of research. The need for a qualitative approach that can be done quickly provides the rationale for Rapid Qualitative Inquiry (RQI). It is my position that RQI provides a useful tool in many situations, but that there are minimum conditions that must be satisfied before the inquiry should be labeled as RQI and that even when RQI is implemented carefully, its limitations should be recognized.

New in the Second Edition

Rapid Qualitative Inquiry: A Field Guide to Team-Based Assessment is a significantly revised and expanded version of *Rapid Assessment Process: An Introduction.* It reflects more than a decade of additional experience on my part, changes in the field of qualitative research, and insights gleaned from the 150-plus users of the first edition who have published or posted their results.

I first began to experiment with, and understand the value of, short-term qualitative fieldwork in the early 1980s, and ever since then my understanding of the basic concepts of rapid research has continued to evolve:

1985 (1) Starting point for understanding a local situation (2) carried out by a multidisciplinary team, (3) at least four days and not more than three weeks, (4) based on information collected in advance, direct observations, and interviews, and (5) based on the assumption that all relevant questions cannot be defined in advance.

1995 (1) System perspective (see ch. 3 "Systems, Soft Systems, and Rich Pictures"), (2) triangulation of data collection, and (3) iterative data collection and analysis.

2001 (1) Intensive teamwork as part of the triangulation of data collection, and (2) intensive teamwork during the iterative process of data analysis and additional data collection.

2014 By definition team based: (1) focused on the insider's perspective, (2) uses multiple sources and triangulation, and (3) uses iterative data collection and analysis.

Among the key features of RQI that will be explored in this edition are an evolving approach to qualitative research that is focused on developing understanding as opposed to finding a single truth, and an exploration of qualitative research in general with special references to what case study research and ethnography contribute.

Whenever possible and appropriate, I have updated references to the latest edition of particular books. However, in a few situations I have used material from both earlier and later editions where I believe the earlier material is not covered in the newer editions. In addition, I have included new references and other resources reflecting the tremendous growth in scholarship concerning qualitative research over the last fifteen years.

Additional changes include:

- a new chapter on getting the insiders' perspective;
- a new chapter on the use of technology;
- a significantly expanded chapter on the RQI Family Tree, providing the context for the development of RQI;
- a final chapter providing a summary organized around key points relating to rigor and some thoughts about the future of RQI;

- expansion of the appendix on teaching RQI skills and attitudes, which now addresses training of practitioners and students, with special attention given to using RQI and Mini-RAP for course requirements involving short-term research projects.

Acknowledgments

Several events and individuals have contributed significantly to my involvement with RQI and to the development and evolution of this book. In 1982, the USAID mission in the Sudan allowed me to experiment with Rapid Appraisal during a weeklong visit to a village in the western Sudan. The second of the two examples of RQI at the beginning of chapter 1 is a brief description of this experience. I returned from this experience convinced of the value of even short-term qualitative fieldwork (see Beebe 1982). I also returned with numerous questions concerning the process that had produced these results. During the next several years there were reports from sites all over the world where people were implementing Rapid Rural Appraisals or similar activities.

In 1985, Khon Kaen University in northeast Thailand and the Ford Foundation sponsored my attendance at the International Conference on Rapid Rural Appraisal at Khon Kaen University (Khon Kaen University 1987). During a small group session at this conference, Terry Grandstaff, M. A. Hamid, Neil Jamieson, and I started the process of identifying and labeling the essential principles of Rapid Appraisal. The conceptual framework used in this book originated in these discussions.

While attending the conference, I met Robert Chambers and had the wonderful opportunity of being with him during a visit to a village close to Khon Kaen University. This visit/fieldwork was designed as a practice session to provide an introduction to Rapid Appraisal. Given the constraints of time and the lack of prior preparation, I was very impressed with Chambers's approach. He started the exchange with a farmer by quickly communicating his respect and defined the role of the farmer as the expert and the role of the visitors as students wanting to learn. The farmer responded by sharing with the team an amazing amount of information in a few minutes. Chambers listened carefully to what the farmer had to say, even while he was observing the environment and using what he saw to direct the conversation. He made a special effort to involve all the other team members in the process.

Since the early 1980s, Chambers has been a leader in defining and popularizing rapid research techniques. His clear writing, devoid of the jargon of the social sciences, and his use of carefully selected phrases that capture the essence of the approach have helped introduce rapid research methods to people worldwide. Since the early 1990s, Chambers has advocated making the process participatory and shifting the focus away from rapid to "relaxed" (Chambers 1999). I owe Chambers a special thanks for his leadership in developing and popularizing rapid research methods.

By 1993 I had reformulated the basic concepts that define Rapid Appraisal and presented them in a paper at a workshop jointly sponsored by the Washington Association of Professional Anthropologists (WAPA) and the applied anthropology program at the American University. In 1995 an expanded version of that paper was published in *Human Organization* (Beebe 1995).

In 1996 I joined the faculty of the Doctoral Program in Leadership Studies at Gonzaga University. Teaching qualitative research forced me to reconsider the relationship between rapid research and qualitative research, including specific approaches to qualitative research such as ethnography and case studies. I started exploring the relationship between the intensive teamwork associated with Rapid Appraisal and the prolonged fieldwork associated with traditional qualitative research.

My students were not shy in pointing out the gaps in my work, and they convinced me of the need to significantly expand the presentation. I owe a special debt to the students in my summer 1999, fall 1999, and summer 2000 qualitative methodology classes, who helped with drafts of the first edition of the book: Dale Abendroth, Elaine Ackerman, Mary Alberts, Una Alderman, Carol Allen, Denise Arnold, Earl Bartmess, Thomas Camm, Nancy Chase, Debra Clemens, Albert Fein, Steve Finch, Craig Hinnenkamp, Rhonda Horobiowski, Kevin Hoyer, Lori Johnson, Janet Katz, Connie Kliewer, Grace Leaf, Kristine Lesperance, Cherisse Luxa, John Lyons, Robert McCann, Susan McIntyre, Matt Mitchell, Barbara Morrison, LaQuitta Moultrie, Michelle O'Neill, David Perry, Jonathan Reams, Marilyn Reilly, Robert Smart, Sandra Smith, Sharon Wessman, Kathryn Whalen, Nancy Bagley, and Terence Young.

A RAP/RQI involving several students from the fall 1999 class provides the other example of student assistance at the beginning of chapter 1. This activity was critical for identifying implementation issues, such as the role of the insider on the team, and is discussed in some detail in chapter 4.

Teaching Advanced Qualitative Research during the spring of 2013 provided an opportunity to rethink the importance of a conceptual framework for qualitative research and the shifts in approaches to qualitative research that have occurred since the first edition. Students and visitors in this class, including Colleen Daniel, John Harper, Russell Horton, Claudine Richardson, and Faith Valente joined me on exploring these topics.

Mitch Allen first suggested I write a book on Rapid Appraisal in 1991. It took me almost ten years of additional experience and a change of careers before I could actually produce the first edition; I appreciate his patience and confidence during that time. The current edition has also benefited from comments and suggestions provided by Rowman & Littlefield's senior acquisitions editor, Leanne Silverman, assistant editor, Andrea O. Kendrick, and the publisher's anonymous reviewers.

A special thank-you to Lauren Angelone, who wrote chapter 7 specifically for this book. Lauren received her PhD from Ohio State University in cultural foundations, technology, and qualitative inquiry. She was one of the first to teach a graduate-level course on technology for qualitative research. As of 2014, she had a consulting business, TransformEd (www.laurenangelone.com) and was working as an adjunct professor at the University of Findlay in Ohio.

I acknowledge the intellectual companionship of my colleagues and their contribution to my thinking about RQI. I especially appreciate the critical help of graduate assistants Patricia Pilcher and Faith Gilbert in cataloging and locating more than one hundred published articles that made reference to the first edition of the book. Their work was used to identify the use of rapid research methods having the characteristics of RQI/RAP in many different fields. I appreciate the helpful comments on this and earlier versions of this work by Andrea Harper, Maria Hizon, Cherisse Luxa, Harold McArthur, Marion McNamara, June Miller, Georgeta Munteah, Perry Phillip, Marleen Ramsey, and Rochelle Rainey. Maria Beebe has been a partner from the beginning and her help is especially appreciated. Obviously, any omissions or errors are my responsibility.

INTRODUCTION

My experience has convinced me that, in a relatively short time, a multidisciplinary research team, including whenever possible **insiders** as well as outsiders, can make significant progress toward understanding a problematic situation from an insider's perspective. My objective here is to convince you that such an approach is possible, to provide you with enough examples and information about specific techniques that you will be willing to experiment with the approach, and to ensure that you recognize its limits.

I believe **Rapid Qualitative Inquiry (RQI)** can produce useful results even when the most important elements of the local situations from the perspectives of the local **participants** are not obvious. Often in these situations, the words the local participants use to define the situation, their categories for dealing with reality, are also not known. If there is no urgent need for an intervention to address the situation, traditional long-term **fieldwork** is a solution. However, my experience has been that there is almost never enough time; that when there *is* time, trained qualitative researchers are not available; and that when both are available, it is exceedingly difficult to convince the decision makers that long-term qualitative research is the best use of resources.

I want to encourage new users to experiment with RQI while helping current users of RQI and related approaches do a better job. I also want to increase confidence in the results of RQI among the decision makers who are its potential clients. Students studying **qualitative research** methods who need to do short research projects, especially students in professional programs, are an important audience for this book. A **Mini-RAP** can provide students with the opportunity to develop skills and attitudes relevant to qualitative research while doing an educational activity based on RQI. Students doing short-term qualitative research activities can be found in graduate and undergraduate courses as diverse as community nursing, advanced agricultural research, rural sociology, forestry, marine management, community development, leadership studies, organizational theory, information systems planning, and urban planning. A second audience are the practitioners facing complex situations where local categories are not known, but who do not have the time or resources for traditional long-term fieldwork. Rapid research approaches have been used for project design, project evaluation, and design of additional research in fields as diverse as wetland evaluation, citywide needs assessment, early childhood care, home ownership patterns among minorities, reproductive health, marketing, and landscape planning (see table 9.1).

RQI uses many of the techniques of qualitative research. These techniques should be familiar to anyone with formal training in anthropology or closely related fields. Ideally, every RQI team will have at least one member with expertise in the assumptions and techniques of qualitative research. However, there will be times when no one on the team has had formal training in qualitative research methodology, or when most team members will be unfamiliar with these techniques. Therefore, I have included brief introductions to some concepts most useful to RQI and suggestions on where additional information can be found. A note of caution—these brief introductions cannot do justice to the richness of qualitative research and the reader is encouraged to seek further information from primary sources. However, even these brief introductions, when combined with the attitude of a seeker, a willingness to listen intently, and genuine respect for others can help users of RQI get started. The specific techniques that are introduced have proven to be especially relevant to RQI. If no one on the team has experience with qualitative research methodology, this should be noted in the report.

My decision to produce a book that can be used in both academic and nonacademic settings will probably ensure that neither camp is completely satisfied with the results.

A Note on Terms

In this second edition I use the phrase **Rapid Qualitative Inquiry (RQI)** to describe a particular type of rapid research methodology. At least some of the characteristics of what I call RQI began appearing in descriptions of research in the late 1970s, under names such as **"Rapid Appraisal," "Rapid Assessment,"** or **"Rapid Rural Appraisal."** More recently, the label **"Rapid Assessment Process" (RAP)** has been applied to a variety of related rapid research approaches. Similar approaches have also been referred to by a variety of names and have been used in numerous settings, in the United States and throughout the world. (These are discussed in chapter 9, "Rapid Research and the RQI Family Tree.")

Although I am reluctant to introduce a new moniker to the mix, there are certain features of rapid research methodology that, taken together, represent a particular and significant type of RAP. RQI can be considered a further refinement of RAP, and in most cases the terms are interchangeable. When I discuss examples of rapid research that were identified as RAP at the time the research was done, but that are also examples of RQI as I define it in this book, usually I will refer to them as RAP/RQI.

Rapid, Qualitative, Inquiry

Rapid Qualitative Inquiry (RQI) is defined as:

> intensive, team-based **qualitative inquiry** with (a) a focus on the insider's or **emic** perspective, (b) using multiple sources and **triangulation**, and (c) using **iterative** data analysis and additional **data collection** to quickly develop a preliminary understanding of a situation.

Each word in the phrase *Rapid Qualitative Inquiry* is useful for further explaining and defining the approach, which is team-based and cannot be implemented by one individual researcher.

For our purposes, *rapid* means a minimum of four or five days and, in most situations, a maximum of about six weeks. RAP/RQI recognizes that there are times when results are needed almost immediately and that the "rapid" production of results involves compromises and requires special attention to methodology if the results are to be meaningful. The time actually required has to be adjusted to the local situation and both the minimum and maximum are offered as guidelines rather than hard rules. Rapid does not mean rushed, and spending too little time or being rushed during the process can reduce RQI to "**research tourism**."

Qualitative refers to a study of things in their natural setting, where the research attempts "to make sense of, or interpret, phenomena in terms of the meaning people bring to them" (Denzin and Lincoln 2011, 3). Creswell (2013, 44) added to Denzin and Lincoln's definition an emphasis on the process of the research with attention to the interpretive nature of inquiry and the need for situating a study within a political, social, and cultural context along with attention to the background of the researcher and how this informs the study. Qualitative research often implies the study of many variables in a few cases. Anthropologists have traditionally used a qualitative research approach to study **cultures**, and anthropology is the academic discipline most closely associated with qualitative research.

Inquiry as used in the phrase *Rapid Qualitative Inquiry*, and consistent with the way the term is used by most people who write about the topic, is a synonym for *research*. However, some observers differentiate by noting that research is expected to contribute to the scientific body of knowledge while inquiry is concerned with practical problem solving. John Dewey is credited with defining inquiry as an investigation into some part of reality with the purpose of creating knowledge for a controlled change. In the field of education, the issue of inquiry versus research has been discussed in terms of teachers doing inquiry for their own benefit as opposed to research being done for the benefit of the community. A more nuanced relationship of inquiry and research in the field of education is suggested by the title of Clarke and Erickson's (2003) book, *Teacher Inquiry: Living the Research in Everyday Practice*. For Reid (2004, 8) inquiry is not carried out with the intention, necessarily, of being made public, but this does not imply that there is any less intellectual rigor in carrying out inquiry than in conducting research. The assumption in *this* book is that the primary purpose of inquiry is not contributing to a broad body of knowledge, but producing understanding that has sufficient rigor to be evaluated publicly and used by others. Results can, but do not necessarily, contribute to a broad body of knowledge. *Inquiry* is also used here as a synonym for assessment and appraisal. I have chosen the phrase *qualitative inquiry* instead of *rapid ethnography* out of respect for those who have helped define **ethnography** and have argued it always requires prolonged fieldwork—but at its heart, RQI is based on ethnography.

Assessment, Appraisal, Mini-RAP

Both *Rapid Assessment* and *Rapid Appraisal* are terms that have been widely used in the kind of research discussed here, often as synonyms. Utarini, Winkvist, and Pelto (2001, 390) suggested assessment "draws attention to a limited or focused scope of information for the purpose of obtaining data to assist in problems solving or evaluation." I did not use the term *appraisal* in the first edition because that term had become so closely associated with development projects, especially projects funded by multilateral donors like the World Bank, that use of the term could lead to confusion and might have limited the potential application of RAP. Many users of RAP did not differentiate between appraisal and assessment and many potential users assumed RAP was an approach used exclusively by international donors in developing countries—an incorrect and unfortunate assumption.

A Mini-RAP is an educational activity and not an approach to research. A Mini-RAP is based on RQI but with limited data collection, usually only two short interviews and a team of two persons. A Mini-RAP can help practitioners master the skills and attitudes needed for doing RQI. Students doing short-term qualitative research as a course requirement will probably do a Mini-RAP but could also do a RQI. The requirements and use of a Mini-RAP are discussed in appendix C.

Field Guides

This second edition is, like the first edition, an introduction, but it is also designed as a field guide.

Field guides have been described as books designed to be brought into the "field" or locale to help the reader identify concepts that are natural occurrences. Contemporary definitions of field guides often make reference to their ability to provide an introduction, make knowledge accessible to individuals without specialized skills, and help readers identify context. They are portable and need not be read from cover to cover in a linear way. A field guide for conducting educational research (California Postsecondary Education Commission 2008) was described as providing guidance and tips based on real-world experience that would allow researchers to apply "deep theory in a hectic and messy world." I especially like their description of their field guide as "a set of variably scaled maps" and their caution that the researcher, like the novice traveler, should "not confuse the map with the territory," since the world of research "is far richer, and messier, than any two-dimensional map." Whether one is an applied researcher working with others to make changes or a student involved in projects to develop qualitative research skills, if the activity is team based, this field guide has been designed as a map rooted in real-world experience.

Organization of the Book

I have organized the book around the three basic concepts of RQI: (1) a focus on the categories used by the insiders to describe local situations; (2) data collection using multiple techniques and triangulation; and (3) analysis using an **iterative process**, where initial analysis is followed by several cycles of additional data collection and more analysis. Chapter 1 discusses two examples of RQI, explores situations where RQI was especially appropriate, and provides an overview of the relationship of specific research techniques to the basic concepts of RQI. Figure 1.1 in chapter 1 is a navigational aid and includes page numbers for quickly locating information about both basic concepts and techniques. Chapter 1 also introduces the important concept of **Appreciative Inquiry**.

Chapter 2 explores the meaning of an insider's perspective and the implication of a focus on insiders' perspectives for research. Chapter 3 deals with data collection and explores the use of multiple data collection techniques and triangulation. Chapter 4 deals with iterative analysis and additional data collection and is divided between (1) an introduction to the iterative process along with a discussion of illustrative techniques associated with the iterative process and (2) a discussion of data analysis and illustrative techniques for data analysis. Chapter 4 also examines issues relating to the team preparation of RQI reports. The materials in chapters 2, 3, and 4 are critical for the teamwork that is the foundation of RQI. The materials in the latter parts of chapters 3 and 4 provide an introduction to research techniques that will be familiar to anyone with training in qualitative research.

Chapter 5 explores the special role of teamwork in RQI and is based on RQIs of the Student Services Division at a community college and state farms in Poland. Chapter 6 examines issues relating to the trustworthiness of RQI's results, notes concerns about the process, and proposes the use of a checklist, the "RAP Sheet," to be attached to RQI reports. The checklist is designed to provide enough information about the research so that a reader can evaluate the results, and to remind the RQI team of issues they should not overlook. Chapter 7 explores technology such as smartphones for improving the team collection and analysis of data and the preparation of reports.

Chapter 8 examines ethical issues including issues concerning the relationship of the RQI team to the sponsoring organization and the issue of **bogus empowerment**. Chapter 9 provides the context for relating RQI to other rapid methods by briefly reviewing the history of RQI. Chapter 9 also includes a list of selected other rapid research studies that is organized around the sectors where these methods have been used. I have only referenced in this book a limited number of the more than 150 studies that I have identified and that used an RQI-based approach. It should be noted that most rapid research studies are used to design interventions or to satisfy course assignments and are not documented in journal articles or even reports that are accessible on the web. A final chapter expands on key concepts related to rigor and considers the future of RQI.

Where appropriate, chapters include the identification of main points and suggestions for additional readings. A significantly expanded appendix C explores ways of teaching RQI to both practitioners and students and explains the special role of the "Mini-RAP." The appendices also include summaries of the two RQI/RAPs that are discussed in the text. A glossary provides definitions of key terms. The first time a term that may not be clear is used it is formatted in bold, indicating its inclusion in the glossary. A list of references and author and subject indexes complete the book.

Beyond the Book

Users of RQI are requested to share their experiences with each other and with me. My e-mail address is beebe@gonzaga.edu. In addition to the website for the book maintained by the publisher, Rowman & Littlefield (www.rowman.com), I have developed a companion website for the Rapid Qualitative Inquiry, http://rapidqualitativeinquiry.com. A Facebook site for the book can be found at http://facebook.com/rapidqualitativeinquiry.

A blog for the book and the topic of rapid qualitative inquiry can be found at http://rapidqualitativeinquiry.blogspot.com/.

Additional Readings

The Essential RQI/RAP Library

In preparing the following list of essential books, I have started with the assumption that expertise in qualitative research methodology is valuable for the RQI team. I have also assumed that there will be times when no one on the team has had prior training in qualitative research methodology and that written materials can provide access to some of the needed expertise. Students in qualitative research courses will find that other books used in their courses also cover the topics in these books. The following books, chosen because I like them and have found them useful, provide both an introduction to the philosophy that underlies qualitative research, to the attitudes necessary to implement it, and details on specific research techniques, including data analysis and preparation of results, relevant to the successful completion of a qualitative research project.

Two Most Useful References for Everyone

Creswell, John W. 2013. *Qualitative inquiry and research design: Choosing among five approaches.* 3rd ed. Thousand Oaks, CA: Sage.

Willis, Jerry W. 2007. *Foundations of qualitative research: Interpretive and critical approaches.* Thousand Oaks, CA: Sage.

Most Useful Reference for Someone with a Very Limited Background in Qualitative Research

Richards, Lyn, and Janice M. Morse. 2013. *Readme first for a user's guide to qualitative methods.* 3rd ed. Thousand Oaks, CA: Sage.

Other Very Useful References

Agar, Michael. 2013. *The lively science: Remodeling human social research.* Minneapolis, MN: Mill City Press.

Angrosino, Michael. 2007. *Doing cultural anthropology: Projects in ethnographic data collection.* 2nd ed. Long Grove, IL: Waveland Press.

Bernard, Russell H. 2011. *Research methods in anthropology: Qualitative and quantitative approaches.* 5th ed. Lanham, MD: AltaMira.
Chambers, Robert. 2008. *Revolutions in development inquiry.* London: Earthscan.
Marshall, Catherine, and Gretchen B. Rossman. 2011. *Designing qualitative research.* 5th ed. Thousand Oaks, CA: Sage.
Miles, Matthew, A. Michael Huberman, and Johnny Saldaña. 2014. *Qualitative data analysis: A methods sourcebook.* 3rd ed. Thousand Oaks, CA: Sage.

Online Resources

The American Anthropological Association's *Anthropology Resources on the Internet*, at http://www.aaanet.org/resources, lists numerous sites with links to anthropology resources. As of 2014 there was an enormous amount of free material available on the Internet at the American Anthropological Association (AAA) and other sites.

The American Anthropological Association offers individuals who do not have access to a college or university library a virtual collection called the *Online Research Library*. This database includes more than five thousand titles, with more than thirty-six hundred in full text. As of 2014, the annual AAA member rate for the Online Library was $29.99 and the nonmember rate was $110 (http://www.aaanet.org/publications/Additional-Journal-Access.cfm).

MIT OpenCourseWare makes the materials used in the teaching of almost all of MIT's subjects available on the web, free of charge. MIT offers many anthropology courses. See http://ocw.mit.edu/courses/anthropology/index.htm for a list.

WikiBooks has free books and courses online that can be downloaded. These resources include a short qualitative research methods course (http://en.wikibooks.org/wiki/Social_Research_Methods/Qualitative_Research) and a general AP Cultural Anthropology course (http://en.wikibooks.org/wiki/Talk:Cultural_Anthropology).

AnthroBase.com (http://www.anthrobase.com/) is a searchable, multilingual database of anthropological texts.

Online Discussion Groups and Access to Selected Journals

Forum: Qualitative Social Research, http://www.qualitative-research.net/index.php/fqs

International Journal of Qualitative Studies in Education, http://www.tandf.co.uk/journals/titles/09518398.asp
Qualitative Inquiry, http://qix.sagepub.com/
The Qualitative Report, http://www.nova.edu/ssss/QR/index.html
Qualitative Research, http://www.sagepub.com/journals/Journal201501?siteId=sageus&prodTypes=Journals&q=qualitative+research
International Journal of Qualitative Methods: http://ejournals.library.ualberta.ca/index.php/IJQM/index
International Review of Qualitative Research, http://www.lcoastpress.com/journal.php?id=8
Field Methods, http://fmx.sagepub.com/
Journal of Contemporary Ethnography, http://jce.sagepub.com/

CHAPTER ONE
THE BASIC CONCEPTS OF RAPID QUALITATIVE INQUIRY (RQI)

Main Points

1. RQI is intensive, team-based qualitative inquiry using triangulation, a focus on the insider's perspective, and iterative data analysis, and additional data collection to quickly develop a preliminary understanding of a situation.
2. In most situations the terms *Rapid Qualitative Inquiry (RQI)* and *Rapid Assessment Process (RAP)* are interchangeable. While I have become convinced that RQI may be a more accurate description and may make the process more accessible than RAP, I continue to use both terms.
3. The phrase *Rapid Qualitative Inquiry* defines the methodology.
4. RQI allows a team of at least two individuals to quickly gain sufficient understanding of a situation to make preliminary decisions for the design and implementation of applied activities or additional research.
5. Results can be produced in as few as four or five days, but often require several weeks.
6. RQI uses the techniques and shares many of the characteristics of traditional qualitative research, but differs in two important ways: (1) more than one researcher is always involved in data collection and teamwork is essential for data collection using multiple techniques and triangulation; and (2) more than one researcher is involved in an iterative approach to data analysis and additional data collection.

7. The intensive teamwork for data collection and analysis is an alternative to prolonged fieldwork and produces results that provide insights into the perspective and worldview of the participants in the local system.
8. RQI is especially appropriate for a variety of situations where qualitative research is needed.
9. RQI can be used for monitoring and evaluation.
10. Sometimes questionnaire/survey research is not an option for initial research because not enough is known to prepare a questionnaire.

Example One: Student Services at a Community College

The new dean at a community college in the Pacific Northwest was made to understand during her interview for the position that there was serious discord in the Student Services Division. After being offered the position, Pat (a pseudonym) was told that she was expected to contribute to a "healing" process. By November of Pat's first year, she realized the rift was greater than she had anticipated but that many of her actions to bring about change were being warmly received by many in the division. Pat knew that there were no easy solutions to the organizational culture issues in the division, and hoped that an examination of this culture could contribute to reconciliation and help individuals refocus on the mission of the division and the college. Pat instinctively knew that the history of the conflicts was too long and the issues too complex for research based on a set of questions to be administered to everyone. She assumed that there would be a very low level of participation in research based on questionnaires, even if appropriate questions could be identified. Pat also knew that the community college did not have the luxury of the extensive time and other resources required for traditional qualitative research. At this point, she requested a Rapid Assessment Process (RAP), an activity that is the same as a Rapid Qualitative Inquiry as described in this book, of the organizational culture of the division. She knew that there was a need to listen to the stories of individuals involved and from these stories to quickly identify some possible interventions.

Staff in the Student Services Division and administrators from the college directly linked to the Student Services Division who contributed to the study, usually by being interviewed, are referred to as participants. Most of the participants' comments were grouped into categories and identified as constraints to the ability of the people involved to do the best job possible. There was a general consensus that performance was not always as good as it should be, and that most people truly wanted to do better. Participants identified numerous, often interrelated, constraints. The six constraints most often identified were: (1) communication; (2) physical space; (3) technology; (4) utilization of people's time, talents, and creativity; (5) increases in the number and complexity of regulations; and (6) inadequate resources.

While there was general agreement that the purpose of working in student services was to provide services to students, there was significant disagreement about the student population and what it meant to serve them. "Students are the reason why I'm here" was a view expressed by almost everyone interviewed. Several individuals, however, objected to the use of the word *service* to describe the relationship with students. One participant suggested that students view services as things they are entitled to and people who provide services as their servants. For some participants, the defining characteristic of service was being "available." This was identified as being an especially important aspect of service for "walk-in traffic." Most participants appeared to recognize a difference between students needing merely "regular" services, such as academic advising, and "high-maintenance" students needing help with issues such as immigration/visa problems and emotional and mental-health problems.

Several participants consistently used the words *customers* or *clients* instead of *students*. There was significant disagreement about the characteristics of the student population and how this had changed over time. Some indicated that "our students have always had special needs," while others pointed to an increase in the number of students with special needs, especially resulting from "overstress in personal academic life choices." A few participants noted their students were less prepared for college now than students in the past had been; one participant said, "Students do not comprehend written communication and have trouble with verbal communication." Another participant noted, "We are doing more hand-holding than we used to do." One person, who had worked at the college for over three decades, said there had been a change in general life experiences that affected preparedness, especially with regard to readiness in a professional/technical arena. Another participant shared a story told to him by one of the technical instructors of a student who was asked to get a Phillips screwdriver from the tool crib, and who returned empty-handed, saying that all he could find was a "Stanley."

Several participants expressed surprise that changes in the characteristics of the student population were even being discussed. "I wasn't aware that there are major changes" in the student population, one participant commented. One participant suggested that perhaps the changes were not in the students but in the staff, saying that the staff had grown older and that maturity had changed perceptions. "I'm not the bleeding-heart liberal I was when I started," this participant added. Some students were described as aggressive and "bullying" in demanding services from staff, "expecting everything to be given to them and believing it's always someone else's fault when things go wrong." Other students were described as lacking sufficient skills to request services they needed and were entitled to. The argument was that, for some students, there was a need for advocacy on their behalf. An individual dealing with special-needs students offered a plea for greater effort to ensure that "no one drops through the cracks" and that his colleagues should realize that this request was not for special advantages, but for a "level playing field."

After the RAP/RQI was over, teams composed of staff from the Student Services Division were organized around the constraints identified by the research and everyone working in the division was asked to participate on at least one team. Funding was made available that teams could apply for to address some of the constraints. The identification of serious differences in how individuals who work in the division understand the terms *students* and *services* resulted in constructive dialogue between the different factions.

To some extent this RQI/RAP is an example of a process that ended up producing solid results despite numerous things going terribly wrong. Too many interviews were scheduled, not all of the interviews were transcribed in time to be used, and it became increasingly difficult to schedule both data collection and data analysis. Some of these issues are discussed in chapter 10 in the sections on "Logistics—Keeping RQI from Becoming SAP," "Schedule Flexibility," and "Interview Notes in Addition to Recording." Flexibility in implementation prevented this from becoming "a bad RAP."

Example Two: Village in the Western Sudan

The soils are sandy, the rain sparse and unpredictable, and the temperatures often above 110°F in this village at the edge of the Sahara in the western Sudan. Here, rural families survive during the better years through a combination of agriculture, livestock, and submission to the will of Allah. Margins for error are extremely slim. The adoption of an inappropriate agricultural intervention could mean ruin for the family and irreversible damage to the environment.

In the early 1980s the U.S. Agency for International Development was providing assistance to Sudan to establish an agricultural research center in western Sudan to help address these issues. The survival strategies of rural households were not fully understood by outsiders. Mohamed el Obeid, a Sudanese agricultural development specialist, and I spent one week in 1982 in a village northwest of al Ubayyid (el Obeid), the provisional capital of North Kordofan (Beebe 1982). My objective was to experiment with the new research methodology called "Rapid Appraisal." The time in the village was amazing. We spent our days talking with groups of farmers or individuals. We conducted interviews about farming practices in the farmers' fields. We had extensive conversations with the village religious leader and the owner of the only shop (one of only two structures made of sun-dried bricks in a village where all the other structures were made of grain stalks and straw). We visited a slightly larger village where farmers could sell grain and other agricultural products. We spent our nights trying to figure out what we had learned and were often joined by men from the village who used our presence as an opportunity to discuss life in general, including the plans of one of the slightly more prosperous inhabitants to take a second wife. The structure imposed by the rapid research methodology allowed us to quickly understand some of the important concerns of the residents of the area. Prior to our visit to this village, the assumption of both Sudanese and American agricultural research scientists was that individual farmers were free to move at any time between gum arabic production (using the nitrogen-fixing tree *Acacia senegal*) and field crops such as sorghum and millet. Attention to their descriptions of crop rotations, especially when this was discussed with farmers in their fields where crops could be observed, suggested that decisions by a farmer's neighbors could be a significant constraint. Gum arabic trees harbor birds, and if a neighbor's large gum arabic trees were too near a farmer's fields, this would prevent the farmer from planting field crops until the neighbor was also ready to cut down his trees in order to sell them for firewood and plant field crops. Subsequent research by Reeves and Frankenberger (1981) on agricultural practices in North Kordofan showed that adjacent fields were often cultivated by farmers who were related and that this also played a role in timing the rotation of a field from gum arabic production to field crops. Recognition of constraints on the decision-making abilities of individual farmers had a significant impact on the approach to farming systems research and extension proposed for the western Sudan.

The Need for the Insider's Perspective

Despite the many kilometers that separate the village in western Sudan and the community college in eastern Washington, the two situations share an important characteristic. Both are complicated situations where initially not enough was known to develop a questionnaire. Only the insiders in each of these situations were in a position to define the elements of their systems and identify those elements that were most relevant to the issues they faced. The insiders in the Sudanese village knew at least intuitively that they could not change their crop rotation pattern without attention to the actions of their neighbors, but it is unlikely that a question could have been formulated in advance that would have elicited this information. Likewise, it is unlikely that a question could have been developed to elicit information on the different ways staff at the community college defined students and services, since there was no reason for outsiders (or even many of the insiders) to think this was an issue. Another characteristic shared by these two situations was that results were needed quickly and that, even if the results had not been needed quickly, there were not sufficient resources for traditional, long-term fieldwork. The approach to research that focuses on getting the insider's perspective is referred to as "qualitative" research or inquiry. In chapter 2, I will return to the relationship between the insider's perspective and qualitative research when discussing **case study** and ethnographic approaches to research (see ch. 2, "Ethnography," "Case Study Research," and "Emic and Etic").

Rapid Qualitative Inquiry and Intensive Team Interaction

Rapid research similar to what was done at the community college in the Pacific Northwest and in the village in the western Sudan provides a way to investigate complicated situations in which issues are not yet well defined and where there is not sufficient time or other resources for long-term ethnographic research or traditional case study research. RQI shares many of the characteristics of qualitative research, especially case study and ethnographic approaches to qualitative research. However, RQI substitutes intensive, team interaction in both the collection and analysis of data for the prolonged fieldwork normally associated with qualitative research. RQI can be expected to produce solid qualitative results that will be different from those produced by longer-term fieldwork. In some cases, intensive team interaction over a short period may produce better results than a lone researcher over a long period. RQI will almost always produce results in a fraction of the time and at less cost than traditional qualitative research.

Basic Concepts and Their Relationship to Research Techniques

RQI uses the techniques and shares many of the characteristics of qualitative research, especially the focus on the need for an insider's perspective. RQI differs in two important ways: (1) more than one researcher is always involved in data collection and teamwork is essential for data collections; (2) researchers use an explicitly iterative approach for data analysis and additional data collection with time allocated for both. The intensive teamwork for both the data collection and the data analysis is an alternative to prolonged fieldwork for producing solid qualitative results. RQI allows a team of at least two individuals to quickly gain sufficient understanding of a situation to make preliminary decisions for the design and implementation of applied activities or additional research. While the length of time suggested for RQI is arbitrary, my experience has convinced me that a minimum of four or five days is usually required for iterative data analysis and additional data collection, an issue I will return to in subsequent chapters (see ch. 6, "Too Little or Too Much Time").

The three basic concepts of RQI define its relationship with traditional qualitative research and allow solid results to be produced quickly. The three basic concepts are:

1. The focus is on getting the insider's perspective.
2. Intensive teamwork is critical for data collection from multiple sources and as part of the triangulation of data collection.
3. Intensive teamwork is critical during the iterative process of data analysis and additional data collection.

Adherence to these concepts can provide a flexible but rigorous approach to the rapid collection and analysis of data.

> RQI is defined by the basic concepts of a focus on the insiders' perspectives, data collection from multiple sources and their triangulation, and iterative analysis and additional data collection, and *not* by the use of specific research techniques.

RQI is defined by its three basic concepts instead of by a specific set of research techniques. While traditionally some research techniques have been associated with **rapid research methods**, these methods are not necessarily required. Specific research techniques for use in a given RQI are chosen from among a wide range of techniques available to qualitative researchers and are chosen based on the specific topic being investigated and the resources available to the team. Specific techniques used in an RQI can vary significantly depending on the situation.

Table 1.1 illustrates the relationship of the basic concepts and illustrative research techniques associated with them. As noted above, the listed research techniques are not the only way of achieving the basic concepts, but are techniques that have been found to work together under some field conditions. Some basic concepts also are techniques. Page numbers in the table point to the location in the text where additional information can be found.

Table 1.1. Basic Concepts and Illustrative Techniques

BASIC CONCEPTS		ILLUSTRATIVE TECHNIQUES	
Concept 1. Focused on the Insiders' Perspectives			
Based on ethnography	32	Use of teams	51
Based on case study research	34	Team work	102
Culture	36	Semistructured team interview	55
Emic and etic	37	Seeking stories and not answers	61
Indigenous knowledge	37		
Variability	38		
Search for understanding	20		
Not searching for a single truth	20		
Concept 2. Multiple Sources of Data and Triangulation			
Information collected in advance	54	Use of teams	51
Semistructured interviews	55	Teams with diversity	51
Observation	54	Small teams	52
Systems and soft systems	68	Teams with insiders whenever possible	52
		Semistructured team interviews	53
		Team observing	54
		Team advanced collection of information	54
		Triangulation as a metaphor	45
		Use of guidelines	60
		Use of an audio recorder	62
		Use of interpreters	62
		Selection of respondents	63
		Focus groups	64
		Comparing and sorting objects	66
		Unobtrusive observations	68
		Rich pictures	70
		Mapping	70
		Field notes	73
		Use of technology	141
Concept 3. Iterative Data Analysis and Additional Data Collection			
Iterative process	81	Team data analysis	112
Data analysis	85	Structuring the research time	82
Data condensation	86	Checking back with participant	84
Additional data collection	82	Team report preparation	84
		Coding	87
		Data display	89
		Conclusion drawing	91
		Verification	93
		Determining how much data is needed	93
		Use of technology	141
		Rapid, not rushed	3

When to Use RQI

I have identified complex situations where the categories and words used by the local people involved in a situation are not known as the type of situations where RQI may be the most appropriate approach. I will try to make the case throughout this book that if results are needed quickly, RQI may be the only choice. RQI is also appropriate for a variety of situations where qualitative research is needed, even if there is no time constraint.

Creswell (2013) identified several situations where qualitative research is especially appropriate that apply to RQI. The first situation described by Creswell is where "a problem or issue needs to be explored" (47). Exploration is described as needed to identify variables "that cannot be easily measured" and to "hear silenced voices" (48). Qualitative research is described as empowering individuals to share their stories and minimizing power relationships that exist between participants and a researcher. Especially relevant to RQI is Creswell's contention that qualitative research is appropriate for understanding an issue by allowing people to tell their stories "unencumbered by what we expect to find" (48).

Willis (2007) identified understanding, and especially understanding in context, as the central objective for qualitative study. Willis's concern is with understanding that can be communicated to others and used for decisions. According to Willis, understanding informs decision makers with ideas that can be tried out and considered as guidelines but not "truths" (121–22). Qualitative study is not the search for universal truths, but attempts to find local truths and better understanding (123).

There is growing recognition of the need for qualitative research in a variety of fields that have traditionally been dominated by quantitative research. Health care and technology come to mind. In June 2006, *Business Week* declared "ethnography" a core competency for business and noted that it provides "a richer understanding than does traditional research" but cautioned about the need to understand requirements for doing it correctly (Ante and Edwards 2006, 102).

Qualitative research, including RQI, is not an appropriate methodology if quantifiable results are needed, such as the percentage of individuals in different categories. RQI may be the appropriate methodology for identifying the most relevant categories, and the most appropriate labels for these categories. However, RQI is usually not the best way to produce numbers.

> RQI may not be appropriate if numbers or percentages are needed.

Using RQI for Monitoring, Evaluation, and Midcourse Corrections

RQI can be used for monitoring and evaluation (see McNall and Foster-Fishman 2007). The identification of specific midcourse corrections during the implementation of an activity is another task for which an RQI may be useful. When ongoing monitoring of an activity suggests problems with implementation and the causes of these problems are not obvious, an RQI team can explore questions as fundamental as whether the local people and the parties responsible for the activity agree on what constitutes success and failure. An RQI approach is especially useful in identifying the unexpected. A report based on a few weeks' work and delivered immediately allows for midcourse corrections. A report prepared by a team in a situation where local people have been full partners increases the chances that recommendations for changes can be implemented and increases the opportunities to implement changes, even before the recommendations have been made formally.

When faced with the limitations of time and resources, the temptation can be to make a very quick visit with the most easily reached local participants. This is sometimes referred to as "research tourism." Another temptation is to do a questionnaire survey even when there is agreement that the most important issues have not yet been identified and that the categories and words with the greatest relevance to the local people are not known. The rationale seems to be that something needs to be done, and that anything, is better than nothing.

Use of RQI When Survey Research Is Not a Good Option

The story of the two neighboring villages in Africa, one where almost all the babies under one year of age were boys and another where almost all were girls, has become part of the folklore of health-care development workers. The villages were identified as a result of a survey of health and mothers' knowledge of health-care practices for children under one. The storyteller usually relates, empathetically, that the study was funded by a major international donor and implemented by a professional researcher who had carefully prepared the questionnaire and trained and supervised the field staff who carried out the interviews. The results were so surprising that local ministry personnel were sent to the villages to investigate. What they reported after their visits to these villages is more interesting than the original "results." The mothers in the first village considered boy babies much more desirable than girl babies and, when asked about their children, tended not to mention the girls. Thus the village appeared to have almost no girls. The mothers in the second village also considered boy babies more desirable than girl babies. However, in this village mothers did not tell strangers about their boys out of fear that if they brought attention to their sons, harm would seek out the boys. Thus the village appeared to have almost no boys. Like so much folklore, it is not possible to identify the source of this story.

Survey research based on questionnaires, a group of written questions to which individuals respond, has been used and misused worldwide. When faced with the need for information about situations, researchers have often tended to do a survey. Survey research has been viewed as reliable, producing similar answers every time the questionnaire is administered, and relatively quick when compared to traditional ethnographic research. My argument is that often survey research is not an option for initial research because not enough is known to prepare the questionnaire. To prepare a questionnaire, you need to be able to identify the relevant elements of a situation, the specific categories that are important to the respondents, and the words they use for these categories. Since a questionnaire cannot identify unanticipated, site-specific relationships, it is limited to validating relationships articulated in advance.

Unless questionnaire survey research is based on the categories and vocabulary of the respondents and the context of the data is understood, the results may not be valid measures of what they purport to measure. Such results can have a degree of **reliability**, since different researchers administering similar questionnaires would likely get the same results and a low degree of **validity**.

An experiment by Stone and Campbell (1984) designed to examine the accuracy of practices, attitudes, and knowledge (PAK) surveys concerning fertility and family planning in Nepal illustrates this. Stone and Campbell hired and trained interviewers to administer the Nepal Fertility Survey to women in three villages. They then cross-checked the information on the survey forms by using other methods, including casual conversations and unstructured interviews. During the survey, 36 percent of the respondents claimed they had not heard of abortion. When Stone and Campbell asked about awareness of abortion, 100 percent knew about abortions and one individual even maintained "that it was inconceivable that someone had not heard of [it]." Stone and Campbell suggest that part of the explanation is that abortion is considered a "religious sin" and that some respondents were "insulted by the question" (1984, 31). When they talked to the respondents who had indicated during the survey that they had not heard of abortion, they found that every respondent had interpreted the question as somewhat threatening. Respondents had interpreted the question on whether they had "heard of abortion" as a question on knowledge of technique or knowledge of who had had an abortion. They also found that every woman who had reported little knowledge of family planning in the survey reported that they had difficulty understanding the questions and had been embarrassed by them.

> Some of them stressed that they were not able to respond to these questions because the interviewers were male. Others said it didn't matter so much that they were male, the problem was that they were strangers. And several women mentioned that they simply could not respond to the questions because other relatives and neighbors were present. (31)

Stone and Campbell identify cultural reinterpretation and problems of context as the factors that influenced the results and noted that these can be problems for research done in the United States as well as research done overseas.

Specific problems have been identified with survey research on sexual behavior, voting, and geographical knowledge that may be relevant to other survey research as well. Clement (1990) suggested that there are two sources of error in research about sexual behavior: invalid answers and the bias resulting from participants in most sexual behaviors studies being volunteers. He noted that the validity of answers depends upon the ability of respondents to remember and their readiness to share the information, and that these are influenced by how the questions are posed (46). He specifically noted problems with cross-cultural research, including research within the United States, but at different universities or with different ethnic groups. Despite increases in the sophistication of survey research methodology and analysis for political/voting surveys, results have become more problematic even while they have become more accurate. More and more individuals are simply refusing to provide answers to pollsters, and increasing numbers identify themselves as "undecided," even when they have decided. Other issues include limited knowledge about issues, changes in when individuals can vote, and the growth of mail-in voting (see Traugott 2005). Surveys on respondents' knowledge of geography have found "astonishing geographic ignorance." Phillips (1993) argued that these findings are the result of asking the wrong question. He noted that the questions are usually based on the categories used by geographers and not the mental images of people whose images are not "map-like." Phillips argued for increased sensitivity to the nature of geographic mental representation as a basis to evaluate geographic knowledge. In all three of these examples, questions that fail to consider the cultural context with specific attention to the definitions and categories of the respondents have produced answers with limited validity.

It is sometimes incorrectly argued that survey research is quicker and can be done with less experienced, less qualified researchers, compared with RQI and other approaches to rapid research. Data collection by survey sometimes requires less time, but data analysis almost always takes more time. Data usually must be coded, entered into a computer, and then analyzed in separate steps and at places removed from the research site. Survey enumerators may have to make fewer independent decisions than a qualitative researcher does, but good survey research cannot be carried out without training and close field supervision. In addition, special training in instrument design and data management ensures that survey research usually does not include local participants as full members on the research team (Chambers 2008, 5–22).

> Beginning research on complex situations with questionnaires may result in the failure to identify important relationships.

RQI may identify the need for questionnaire survey research to supplement its results. As noted above, RQI cannot provide information on the percentages of respondents in different categories, and this information can be critical. Survey research can provide this type of information. However, RQI can provide the categories, vocabulary, and context necessary for the preparation of the questionnaire. The argument here is not against using questionnaire surveys, but against using them as the first step for trying to understand complex situations before local categories are known. In addition, an RQI may be a better starting point for some research because of its ability to discover relationships within the situation that may not have been anticipated. The use of techniques associated with RQI does not guarantee success in identifying important relationships, but initial research on complex situations based on a questionnaire often ensures that they will be missed.

RQI and Appreciative Inquiry

RQI can greatly benefit from an Appreciative Inquiry approach, but does not require it. Appreciative Inquiry focuses on identifying what is going well, determining the conditions that make excellence possible, and encouraging those conditions within the organizational culture. Hammond (2013) described Appreciative Inquiry as a way of thinking, seeing, and acting for purposeful change in organizations. She suggested that the focus for traditional organizational development consultants is looking for problems and that looking for problems ensures they are found and made bigger. Table 1.2 compares tradition in approaches to research with Appreciative Inquiry.

The assumptions of Appreciative Inquiry are:

1. In every society, organization, or group, something works.
2. What we focus on becomes our reality.

Table 1.2. Traditional versus Appreciative Inquiry

Traditional	*Appreciative Inquiry*
Define the problem.	Search for solutions that already exist.
Fix what is broken.	Amplify what is working.
Focus on decay.	Focus on life-giving forces.
What problems are you having?	What is working well around here?

3. Reality is created in the moment and there are multiple realities.
4. The act of asking questions of an organization or group influences the group in some way.
5. People have more confidence and comfort to journey to the future (the unknown) when they carry forward parts of the past (the known).
6. If we carry parts of the past forward, they should be what is best about the past.
7. It is important to value differences.
8. The language we use creates our reality (Hammond 2013, 14–15).

Appreciative Inquiry is based on engaging the entire system in a discussion of what works, with analysis focused on discovering what could be. The future is "envisioned" through an analysis of the past and the best of the past is maintained and stretched into the future (Hammond 2013).

The Need for Caution Concerning the Use of RQI

Robert Chambers's observation concerning Rapid Appraisal also applies to RQI, in that there is a danger it "could be over-sold, too rapidly adopted, badly done, and then discredited, to suffer an undeserved, premature burial as has occurred with other innovative research approaches" (1991, 531). As already noted, when numerical data is needed, RQI by itself will probably be an inappropriate methodology, but it might contribute to the design of a survey questionnaire to collect numerical data. When situations that need to be investigated are especially complex or cover an entire cycle, such as a growing season or a school year, long-term qualitative research may be necessary. There are situations where the research is part of the intervention to bring about change and where extended time in the field is critical for building or restoring trust (see Spoon and Arnold 2012). There may be situations where it is culturally inappropriate for a team of researchers to interview an individual. Other valid reasons for concern about RQI include spending too little time on the activity, failure to consider the political and economic context, problems with team composition, choice of respondents and informants, and a failure to recognize a difference in power between the team and the **local community**. To date there has been a general lack of confirmation of RQI and other rapid qualitative research findings. RQI, along with other qualitative research methods, lacks credibility with some funding agencies, while other funding agencies have very unrealistic expectations about what RQI can accomplish and sometimes pressure researchers to do RQIs in inappropriate situations. These issues will be discussed in more detail in chapter 6 (see ch. 6, "Problems with Credibility").

> There are numerous situations where RQI is inappropriate, including when numbers are needed, when long-term research is part of an intervention, and when only one researcher can be present for data collection.

In the next chapter, chapter 2, I will discuss the first of the three basic concepts, the one relating to the focus on understanding the insiders' perspectives. In chapter 3 I will explore the closely related topics of data collection, triangulation, and intensive teamwork and their relationship to beginning the process of understanding the insider's perspective. Because of their relevance to RQI, ethnography and case study approaches to qualitative research will be discussed in chapter 2. However, since this is not intended to be a book about ethnography, case study, and qualitative research in general, issues can only be introduced and the reader is encouraged to refer to the additional readings listed at the end of the chapters for more information. The specific techniques that are introduced are those that have proven to be most relevant to RQI. Most of the techniques are designed to help facilitate the telling of stories as opposed to the eliciting of answers. Others are designed to ensure that the RQI team records data in ways that will make the data useful and easier to analyze. A specific RQI may use only a few of these techniques and may use other techniques that are not covered. One of the strengths of RQI is that it is not based on the use of a specific list of techniques.

Additional Readings

The readings listed below are some of the most-often-cited references dealing with rapid research methods. Even though Scrimshaw and Gleason

(1992) focus on health programs, material in this book will be useful to researchers from a variety of fields. A selected list of articles dealing with Appreciative Inquiry, can be found at http://appreciativeinquiry.case.edu/intro/classics.cfm.

Chambers, Robert. 2008. *Revolutions in development inquiry*. London: Earthscan.

———. 1991. Shortcut and participatory methods for gaining social information for projects. In *Putting people first: Sociological variables in rural development*, 2nd ed., edited by M. M. Cernea, 515–37. Washington, DC: Oxford University Press, World Bank.

Cooperrider, David L., and Diana Kaplin Whitney. 2005. *Appreciative inquiry: A positive revolution in change*. San Francisco, CA: Berrett-Koehler.

Hammond, Sue A. 2013. *The thin book of appreciative inquiry*. 3rd ed. Plano, TX: Thin Books.

Khon Kaen University. 1987. *Proceedings of the 1985 International Conference on Rapid Rural Appraisal*. Khon Kaen, Thailand: Rural Systems Research and Farming Systems Research Projects.

Kumar, Krishna. 1993. *Rapid appraisal methods*. Washington, DC: World Bank.

Scrimshaw, Nevin S., and Gary R. Gleason. 1992. *Rapid assessment procedures: Qualitative methodologies for planning and evaluation of health related programmes*. Boston: International Nutrition Foundation for Developing Countries. Online version http://archive.unu.edu/unupress/food2/UIN08E/UIN08E00.HTM

Van Willigen, John, and Timothy Finan. 1991. *Soundings: Rapid and reliable research methods for practicing anthropologists*. Washington, DC: American Anthropological Association.

CHAPTER TWO

GETTING THE INSIDERS' PERSPECTIVES

Main Points

1. RQI and qualitative research share the objective of eliciting the insider's perspectives.
2. RQI involves a trade-off between the number of cases investigated and the number of variables examined for each case. Qualitative research and RQI are based on a few cases and many variables.
3. Results are often ambiguous. The RQI team should practice the mantra "I can live with ambiguity."
4. "Ethnography is based on learning from people as opposed to studying people" (Spradley and McCurdy 1972, 12).
5. The purpose of ethnography is to "describe what the people . . . do, and the meaning they ascribe to what they do" (Wolcott 2008, 72).
6. "Qualitative research, then, has the aim of understanding experience as nearly as possible as its participants feel it or live it" (Sherman and Webb 1988, 7).
7. The two principal approaches to collecting ethnographic data are participant observation and semistructured interviews.
8. The intent of case study research is to develop an in-depth understanding of a single case or issues across multiple cases.
9. RQI, like case study research, has a tendency to attempt to answer questions that are too broad.

10. An emic approach attempts to understand the categories the local people use for dividing up their reality and identifying the terms they use for these categories.
11. Recognition of variability can lead to interventions that provide people with options and that respect their ability to choose among them.

Qualitative Research and Developing Understanding Based on the Insiders' Perspectives

Rapid Qualitative Inquiry is based on what the participants in the situation under investigation believe are the critical elements, the relative importance of these elements, and how they relate to each other. RQI is designed to elicit the insider's perspective, a central objective for all qualitative research. RQI and traditional qualitative research share many characteristics and specific research techniques. The most significant difference between RQI and traditional qualitative research is RQI's use of intensive teamwork to substitute for the prolonged fieldwork associated with most approaches to qualitative research. RQI does not reject or abandon the traditional methods and techniques of the social sciences, but provides for ways to complement and enrich them. A brief examination of qualitative research provides the context for RQI.

Qualitative research, and especially ethnography and case study approaches to research, (1) helps define for RQI what it means to get at the insider's perspective and (2) provides research techniques.

Sherman and Webb (1988) identified five characteristics shared by all qualitative research:

1. Events can be understood adequately only if they are seen in context.
2. Nothing is predefined or taken for granted.

3. Participants need to speak for themselves. For participants to be able to speak for themselves, there is a need for an interactive process between the persons studied and the researcher.
4. The aim of qualitative research is to understand the whole experience.
5. There is not one method, and choices are made based on appropriateness (5–8).

Sherman and Webb focused on the importance of the perceptions of the participants in their much-quoted definition: "Qualitative research, then, has the aim of understanding experience as nearly as possible as its participants feel it or live it" (7). Denzin and Lincoln (2011) also emphasized the perspective of the people studied in their statement that "qualitative researchers study things in their natural settings, attempting to make sense of or interpret phenomena in terms of the meaning people bring to them" (3). Miles, Huberman, and Saldaña (2014) noted that the researchers attempt to capture the perceptions "from the inside," through a process of "deep attentiveness, of empathetic understanding," and that this requires researchers to suspend preconceptions about the topics under investigation (9). They argued that "a main task is to describe the ways people in particular settings come to understand, account for, take action, and otherwise manage their day-to-day situations" (9). Creswell's 2008 definition of qualitative research referred to "addressing the meaning individuals or groups ascribe to a social or human problem" (2008, 4). Creswell (2013, 14) summarized common characteristics of qualitative research described by several leading authors on the subject. He suggested that these authors agree that qualitative research is conducted in a natural setting, involves using multiple methods, and involves an emergent and evolving design. Creswell noted that a trend in qualitative research has been "closer attention to the interpretive nature of inquiry and situating the study within the political, social, and cultural context of the researcher" (45). Creswell noted his definition of qualitative research has a "strong orientation towards the impact of qualitative research and its ability to transform the world" (44).

Two other characteristics of qualitative research are especially relevant to RQI and need to be embraced by RQI team members. First, all social science research involves a trade-off between the number of cases investigated and the number of variables examined for each case. Qualitative research is based on a few cases and many variables, whereas quantitative research works with a few variables and many cases (Ragin 1987 as cited in Creswell 1998, 15–16). When the objective is to understand the perspective of the insiders with special attention to understanding their categories, fewer cases covered in depth are preferred to more cases covered in less depth. Second, qualitative research assumes there will be ambiguity. Qualitative researchers cannot know what they will learn and must be able to tolerate uncertainty (Ely 1991, 135). They also need to recognize that what they find may not have one clear meaning. Results are often ambiguous. The RQI team should practice the mantra "I can live with ambiguity" until they believe it.

The RQI mantra: "I can live with ambiguity."

Case Study and Ethnography as Approaches to Qualitative Research

The numerous approaches to qualitative research (see Creswell 2013) reflect different disciplines and are based on different procedures for data collection and analysis, and the production of reports. It is not always clear what differentiates one approach from another and there is little agreement on even the number of different approaches. Studies can be based on a combination of different approaches. It is my contention that case study and ethnography are approaches to qualitative research that are especially relevant to RQI.

Ethnography

While I especially like the statement by Spradley and McCurdy (1972) from more than forty years ago that "ethnography is based on learning from people as opposed to studying people" (12), it does not differentiate ethnography from other qualitative research. Miles et al.

(2014, 8) provided a comprehensive description of ethnography as staying close to the naturalist form of inquiry and involving "(a) extended contact within a given community; (b) concern for mundane, day-to-day events as well as for unusual ones; (c) direct or indirect participation in local activities . . . ; (d) a focus on individuals' perspectives and interpretations of their world; (e) relatively little pre-structured instrumentation; . . . and (f) more purposeful observation." Creswell (2013) defined ethnography as describing and interpreting the shared and learned patterns of values, behaviors, beliefs, and language of a culture-sharing group (90). For Creswell (2013) the defining characteristics of ethnography include a focus on developing a description of the culture, looking for patterns of social behavior, using theory to focus the research, and conducting extensive fieldwork (91–92). Creswell suggested interviewing and **participant observations** as important elements of fieldwork and the concepts of emic and **etic** as important for analysis of culture (see "Emic and Etic"). Wolcott described ethnography as a perspective rather than a means of data collection (Wolcott 2008). The purpose of ethnography, according to Wolcott, is to "describe what the people in some particular place or status ordinarily do, and the meaning they ascribe to what they do" while drawing attention "to regularities that implicate cultural process" (2008, 72–73). Wolcott's definition of ethnography as a study of a cultural group or individuals within the group based primarily on observation introduces the important concept of cultural group. The extensive references to cultural groups and culture help differentiate ethnography from other qualitative research.

The two principal approaches to collecting ethnographic data are participant observation and **semistructured interviews**. When conducting ethnographic research, the researcher observes a situation and listens to informants and records their voices with the intent of generating a "cultural portrait" of the situation (Wolcott 2008).

Almost all descriptions of ethnography refer to a requirement for prolonged periods in the field, often defined as twelve months or more (Bernard 2011; Creswell 2013; Miles et al. 2014; Wolcott 1995, 2005). This topic will be discussed in more detail in chapter 6, when I consider the arguments by some anthropologists that an RQI is too quick to be ethnographic research. Despite the widespread identification of ethnography with prolonged fieldwork, there are occasional references to the possibility that ethnography can be done in less time. In addition to the researchers discussed in chapter 9 associated with rapid ethnographic research (Robert Redfield and Sol Tax, James Spradley and David McCurdy, and Penn Handwerker), Margaret Mead is described as being neither bashful nor apologetic about the short duration of her fieldwork experiences (Wolcott 1995, 78).

There are some situations where prolonged fieldwork is required to gain access and build rapport. There are other situations involving events that unfold slowly, like agricultural cycles or the merger of organizations, where prolonged fieldwork is necessary. However, for most situations, RQI should produce a sufficiently rich understanding of the insider's perspective to initiate interventions that need to be started quickly or to design additional research. RQI is not used to do an ethnography. Rather RQI uses some of the concepts of ethnography to develop the beginning of an understanding of a situation from the insider's perspective.

Case Study Research

Creswell (2013, 97) suggested that while the intent of ethnography is to explore how a culture works, the intent of case study research is to develop an in-depth understanding of a single case or explore issues or problems across cases. According to Creswell a case study is a qualitative approach for exploring a real-life contemporary bounded system (a case) or multiple bounded systems (cases). Cases are explored through detailed, in-depth data collection involving multiple sources of information such as observations, interviews, audiovisual material, and documents and reports. While several researchers have written about the use of case study, including Creswell (2013), Denzin and Lincoln (2005), Merriam (1998), and Woodside (2010), I have come to depend on Stake (1995, 2006) and Yin (2014) as authorities. All of the approaches share the goal of ensuring that a topic is well explored. Yin defined a case study as "an empirical enquiry that: investigates a contemporary phenomenon within its real-life context, especially when the boundaries between phenomenon and context are not clearly evident" (Yin 1994, 13). For Stake (2006, 8) the focus of case study is on "the particular and the situational" and the term *case study* applies to both a process of inquiry and the product of that inquiry.

Baxter and Jack (2008) added to the definition of case study research a focus on exploring a phenomenon within its context and the use of a variety of data sources. "This ensures that the issue is not explored through one lens, but rather a variety of lenses which allows for multiple facets of the phenomenon to be revealed and understood" (544).

Baxter and Jack (2008, 545) noted that both Stake (1995) and Yin (2003) base their approach to case study on a constructivist paradigm. For constructivists truth is relative and is dependent on one's perspective. Reality is socially constructed. An advantage of a constructivist approach is the assumption that the research and the participants will collaborate and the participants will be encouraged to tell their stories and describe their views of reality.

According to Creswell (2013, 98) the first step in a case study is to identify a specific case to be investigated. Creswell further noted that a case can be an individual, a small group, an organization, a community, a relationship, a decision process, or a specific project. One of the problems with a case study approach identified by Baxter and Jack (2008, 546), a problem that is especially relevant to RQI, is the tendency to attempt to answer a question that is too broad. Both Yin (2014) and Stake (2006) suggested the use of boundaries to focus studies. Yin (2014, 33–34) referred to binding the case and suggested, as examples, the persons to be included, geographic area, and the beginning and ending time. Stake (2006, 3) noted it can be difficult "to draw a line marking where the case ends and where its environment begins, but boundaries, contexts, and experience are useful concepts." Stake suggested boundaries based on time and activity. Creswell (2013) suggested boundaries for case study based on time and place. There is general consensus that a case study report needs to include a description of the case and a finding section of the themes or issues that the researcher has uncovered. The results of case study research have been called "assertions" by Stake (2006), "explanations" by Yin (2014), and "lessons learned" by Creswell (2013).

Stake (as cited by Creswell 2013, 96) rejected calling case study a methodology and instead advocated calling it an approach or choice of what is to be studied. The focus on process as opposed to procedures is another area where case study has implications for RQI.

Myers (2009, 83) provided a list of factors that make for an exemplary case study: the case study must be "interesting," the case study must display sufficient evidence, the case study must be "complete," the case study must consider alternative perspectives, and the case study should be written in an engaging manner. He also identified the plausibility of the story as an indicator of quality. Myers provided an important reminder that one case study may be sufficient.

Key Concepts of Qualitative Research Relevant to RQI

Culture

The generation of a cultural portrait has been identified as one of the purposes for qualitative research. The words *cultural* and *culture* have special meaning when used by qualitative researchers, and these meanings have implications for RQI. Some social scientists have defined culture as nearly everything that has been learned or produced by a group of people. A more limited definition first proposed by Spradley and McCurdy (1972) and slightly revised in McCurdy, Spradley, and Shandy (2005) restricts the concept of culture to "the knowledge that is learned and shared and that people use to generate behavior and interpret experience" (5). McCurdy et al. noted that this definition has consequences for researchers because one cannot see knowledge—"it is located in people's heads"—and thus culture has to be inferred from the behavior and objects people in a culture have produced. McCurdy et al. noted that speech is a kind of behavior and can be used to deduce cultural knowledge. Spradley and McCurdy proposed that the concept of culture "shifts the focus of research from the perspective of the ethnographer as an outsider to a discoverer of the insider's point of view" (1972, 9). The task of the researcher is not to discover everything that a group of people has learned, but to discover the insider's perspective on the shared knowledge people use for social behavior. Since culture is learned, and since we all learn different things, no two people can have identical cultures. This also means that any one person will have different cultural configurations at different points in his or her life. The fact that culture is learned and shared, but that each individual participates in different groups where she/he learns different things, means that everyone participates in many cultures. Finally, all cultures are dynamic. Disagreements among local people as to what are the significant categories, or even the words they use for these categories, should be expected.

Emic and Etic

One of the goals for qualitative research, including case study, ethnography, and RQI, is to understand the categories the local people use for dividing up their reality and identifying the terms they use for these categories. Spradley (1979, 102–3) urged researchers to distinguish between what is observed and what the researcher thinks about what was observed. Spradley warned against "the temptation . . . to create order out of other cultures by imposing your own categories." The categories used by the local people are referred to as "emic" and the categories used by the researcher, if he or she is an outsider, are referred to as "etic." When I was a graduate student, these terms were illustrated with the following example. The first situation illustrates an etic approach. If I am interested in the way a group describes the colors of objects, I could start by pointing to an object that is dark green and then inquire about the word they use. I have started with my classification of colors, where there is a discrete category for dark green, and it is unlikely I will discover that their categories for colors, the way they divide colors, is different from my categories. An emic approach would be to suspend for a moment my classification system and to start by trying to understand their categories. I might present a color wheel where there are no divisions between the different shades and ask the respondent to provide the boundaries and to then provide labels. I could also start with a color photograph or natural objects and collect information on colors. Using this approach, I would discover that several groups in southern Sudan have only one category that includes dark green and black and only use one term to describe them. Closer to home, the emic system used by students in an American high school will be found to have numerous categories for their classmates, including categories of "wannabes." It quickly becomes obvious that a given student could fit into numerous categories and that different students, even though they share in a similar culture, will identify slightly different classification systems.

The following summary of an "emic" approach is based on Pelto and Pelto (1978, 62):

1. Primary method is interviewing, in-depth and in the local language.

2. Intent is to seek the categories of meaning, as nearly as possible in the ways the locals define things.

3. The people's definitions of meaning, their idea systems, are seen as the most important causes or explanations of behavior.
4. Systems and patterns are identified through logical analysis.
5. The methodology begins with particular observations and uses them as a foundation for understanding, since research cannot proceed until the local categories of meaning have been discovered.

The insider's perspective is an emic perspective.

Two points of caution are necessary. No one waits to identify all of the local categories before beginning to discuss issues or ask questions. The assumption has to be that some of the outsiders' definitions are useful. The goal is to be sensitive to the possibility that the categories of the insiders may not be the same as the categories of the outsiders and, where appropriate, to try to elicit the emic categories. The second point is that communication about the local situation may require that emic categories be translated into etic ones that can be understood. An excellent study on firewood/energy use in northern Sudan made extensive use of emic categories. The local people use a "donkey" load as their unit of measurement for firewood. The author of the study also referred to donkey loads in her report without defining the term, and consequently some of her most important research findings were of limited value to the rest of the world.

Indigenous Knowledge

The beginning point for understanding complex local situations has to be the understanding of the local participants. The goal of RQI is to construct a model of the local situation consistent with the way local people understand it. Doing so usually means trying to use local categories for dividing and describing reality. Using indigenous knowledge involves seeking agreement on the most important problems or constraints faced by the local participants (Galt 1987). Indigenous knowledge of local systems cannot capture the totality of these situations and there will always be areas where local understanding of reality is limited. The involvement of outsiders on the RQI team can help move the understanding of the situation beyond that possible by the local participants. At the same time, the outsiders will need the continuous input of the local participants to avoid serious errors understanding the situation (Galt 1987).

Variability

One of the tasks of research is to recognize and seek out variability and not to focus exclusively on the "average." In many situations, the average student, small businessperson, substance farmer, or rural health-care administrator exists only as an artifact of statistics. If the average weight is defined as a ten-pound range for children of a specific age, it might be found that 30 percent of children are average. If the average height is defined as a five-inch range, it might also be found that 30 percent of children are average. However, if both height and weight are used, the number of children who are average will be far less than 30 percent, because not everyone with an "average" height will also have an "average" weight. Each time, an additional variable is used to define the average, fewer and fewer actual cases of the "average" can be found. Eventually, if enough variables are used to define the average, almost no one is left who is average. In many situations, variability and distributions of characteristics are more important than the "average." Qualitative research approaches that focus on the average or are implemented with inadequate time in the field are especially prone to ignore variability. Ignoring variability can be dangerous when it results in the development of interventions that the "average" recipients are expected passively to adopt. Recognition of variability can be an important beginning point for developing interventions that provide people with expanded options and that respect their ability to choose among them (Beebe 1994).

Additional Readings

Chapters 2 and 3 share a common section on additional reading. Three entries from the Essential Rapid Qualitative Inquiry/RAP Library (see the introduction) are included here as well. Richards and Morse's (2013) *Readme First for a User's Guide to Qualitative Methods* is strongly recommended as an introduction for individuals with little or no background in qualitative methods. Alternative introductions to social science research designed for the practitioner include Wadsworth's (2011) *Do It Yourself Social Science* and Guest, Namey, and Mitchell's (2013) *Collecting Qualitative Data: A Field Manual for Applied Research.* Creswell's (2013) *Qualitative Inquiry and Research Design: Choosing among Five Approaches* provides an introduction to the different aspects of qualitative research including examples and lists of key references. Separate sections cover case study and ethnography. Creswell's Analytical Table of Contents by Traditions is especially useful. Willis's (2007) *Foundations of Qualitative Research: Interpretive and Critical Approaches* introduces key theoretical and epistemological concepts in an accessible style with numerous examples. Bernard's (2011) *Research Methods in Anthropology: Qualitative and Quantitative Approaches* provides in-depth explanations of specific techniques. Yin's (2010) *Qualitative Research from Start to Finish* is both practical and scholarly and could be an alternative to Creswell. Case study key references include Stake's (2005) "Qualitative Case Studies," Stake's (1995) *The Art of Case Study Research*, and Yin's (2014) *Case Study Research Design and Methods.* An updated alternative to Stake (1995) is Stake's (2006) *Multiple Case Study Analysis.* Ethnography key references include Fetterman's (2010) *Ethnography: Step-by-Step* and Wolcott's (2008) *Ethnography: A Way of Seeing.*

Bernard, H. Russell 2011. *Research methods in anthropology: Qualitative and quantitative approaches.* 5th ed. Lanham, MD: AltaMira.

Creswell, John W. 2013. *Qualitative inquiry and research design: Choosing among five approaches.* 3rd ed. Los Angeles, CA: Sage.

Ervin, Alexander M. 2005. *Applied anthropology: Tools and perspectives for contemporary practice.* 2nd ed. Boston: Pearson/Allyn and Bacon.

Fetterman, David M. 2010. *Ethnography: Step-by-step.* 3rd ed. Los Angeles, CA: Sage.

Guest, Greg, Emily E. Namey, and Marilyn L. Mitchell. 2013. *Collecting qualitative data: A field manual for applied research.* Los Angeles, CA: Sage.

Kvale, Steinar, and Svend Brinkmann. 2009. *InterViews: Learning the craft of qualitative research interviewing.* 2nd [rev.] ed. Los Angeles, CA: Sage.

Richards, Lyn, and Janice M. Morse. 2013. *Readme first for a user's guide to qualitative methods.* 3rd ed. Los Angeles, CA: Sage.

Stake, Robert E. 2006. *Multiple case study analysis.* New York: Guilford Press.

———. 2005, Qualitative case studies. In *The Sage handbook of qualitative research*, 3rd ed., edited by N. K Denzin and Y. S. Lincoln, 443–66. Thousand Oaks, CA: Sage.

———. 1995. *The art of case study research*. Thousand Oaks, CA: Sage.

Willis, Jerry W. 2007. *Foundations of qualitative research: Interpretive and critical approaches*. Thousand Oaks, CA: Sage.

Wadsworth, Y. 2011. *Do it yourself social science*. 3rd ed. Walnut Creek, CA: Left Coast Press.

Wolcott, Harry F. 2008. *Ethnography: A way of seeing*. 2nd ed. Lanham, MD: AltaMira.

Yin, Robert K. 2014. *Case study research: Design and methods*. 5th ed. Thousand Oaks, CA: Sage.

———. 2010. *Qualitative research from start to finish*. New York: Guilford Press.

CHAPTER THREE

DATA COLLECTION—MULTIPLE DATA SOURCES AND TRIANGULATION

Main Points

1. The synergy resulting from triangulation based on intensive teamwork is absolutely crucial for RQI.
2. Intensive team interaction among RQI team members is an element of triangulation (along with team interaction for the iterative analysis and additional collection of data) and permits the RQI team to reduce significantly their time in the field.
3. The strength of RQI results from interaction among the RQI team members and the ability of the team to carry out triangulation at a rapid rate.
4. By definition, RQI cannot be done by one person. Whenever possible teams should be composed of a mix of insiders from and outsiders to the situation being investigated.
5. Intensive team interviewing is not based on sequential interviewing by members of the RQI team, but on joint interviewing.
6. RQI and ethnography share many characteristics and specific research techniques.
7. Semistructured interviews based on guidelines are the key to RQI. The most important way of learning about local conditions is to ask local people.
8. The goal is to get people to talk on a subject and not just answer questions.

9. More than twenty techniques associated with qualitative research and RQI are introduced in this chapter. To begin experimenting with RQI, all you need to remember is that your goal is to talk with people and to get them to tell their stories, as opposed to answering your questions.

Use of Multiple Sources of Data and Triangulation

An Example

The individual researcher sits across the desk from the director of the not-for-profit organization and asks her for details about the recent personnel crisis that appears to threaten the organization. The researcher is conscientious about trying to combine information he has collected in advance about the personnel issue with observations of the office as ways of informing the interview of the director, but problems making the tape recorder work distract him. About the only information he takes from this situation is what the director chooses to share with him during the interview.

Imagine this same scene, but with a team of researchers. This time a human relations/personnel specialist and a financial specialist join the lead researcher. One of the team members who will not be starting the interview takes care of the tape recorder. The lead researcher notices two fairly large stacks of unopened mail on a side table and begins the discussion with a reference to them. The comments by the director about her inability to keep up with events remind the financial expert of the failure of the organization to complete a scheduled audit and he now joins the discussion. It becomes obvious that the personnel issues are not isolated. When the conversation turns to personnel, the human relations expert uses her expertise to ask about the rationale for some recent policy changes. She is sensitive to the feelings of the director and takes time to ensure that there is no appearance that this is "**tag-team interviewing**." The lead researcher notices that the human relations expert has not followed up on a potentially important comment by the director, and seeks clarification. By now you should get the picture. The same amount of time is spent on the interview, but the results are far richer than what the single researcher could have achieved. The results are even richer than what the three researchers could have achieved if each had separately interviewed the director.

Triangulation, Multiple Perspectives, and the Insider's Perspective

Before exploring the relationship between RQI and triangulation based on intensive teamwork, it is necessary to briefly introduce the concept of triangulation and to discuss some of the issues with the use of the term. The term *triangulation* comes from navigation and physical surveying and describes an operation for finding a position or location by means of bearings from two known fixed points. For someone wanting to locate their position on a map, a single landmark can only provide the information that they are situated somewhere along a line in a particular direction from that landmark. With two landmarks, however, their position can be pinpointed by taking bearings on both. They are at the point on the map where the two lines cross. Sometimes the exact location of these landmarks is not known, and triangulation can only identify a position in relation to the known points and not an exact point.

Triangulation has been used as a **metaphor** by social scientists for the use of data from different sources, the use of several different researchers, the use of multiple perspectives to interpret a single set of data, and the use of multiple methods to study a single problem. Even though the origin of triangulation refers to the use of three points, the term should not be taken literally; it is not limited to processes involving three sources of data such as researchers, perspectives, or methods (Janesick 1994, 215). The specific research techniques associated with triangulation are discussed later in this chapter. Intensive teamwork can increase the power of triangulation exponentially. Special requirements for successful teamwork, including a consideration of roles in and leadership of the team, are discussed in chapter 5.

Triangulation is seen as the combination of methodologies for the study of a particular phenomenon (Flick 1992, 176) and it is used to test one source against another. Fetterman (1998) noted that triangulation defined this way "always improves the quality of data and the accuracy of ethnographic findings" and is at the heart of ethnographic validity (89). The assumption is that the researcher is searching for the convergence of at least two pieces of data (Ely 1991, 97). It is further assumed that triangulation will result in better propositions.

When triangulation is focused exclusively on the convergence of evidence, Ely (1991, 98) suggested, the other outcomes, inconsistency and contradiction, may be missed, although these are of greater value. Ely noted that the data that is inconsistent or contradictory are sometimes called negative cases. Negative case analysis is the search for evidence that does not fit into the emergent findings and that can lead to a reexamination of the findings (98).

Mathison is among the social scientists who suggest that convergence should not be expected. She noted that social phenomena are always complex and that triangulation should not be expected to provide a "clear path to a singular view of what is the case" (Mathison 1988, 15). For an increasing number of social scientists, triangulation is not a tool for validation, but an alternative to validation (Denzin and Lincoln 1994, 2). Since, according to Denzin, objective reality will never be captured, the goal of triangulation is the creation of fully grounded interpretive research (cited in Flick 1992, 180). For these social scientists, triangulation offers different perspectives and gives access to different versions of the phenomenon that is studied. Research is not intended to identify one reality against which results can be verified or falsified, but deals with different versions of the world (Flick 1992). Denzin and Lincoln argued that "the combination of multiple methods, empirical materials, perspectives, and observers in a single study is best understood then, as a strategy that adds rigor, breadth, and depth to any investigation" (1994, 2). Similar views were expressed by Flick (1992) and Fielding and Fielding (1986). Flick referred to the ability of triangulation to add "breadth and depth" without "artificial objectivation of the subject under study" (194). Fielding and Fielding suggested triangulation based on the addition of carefully and purposefully chosen methods is done to add "breadth or depth to our analysis, not for the purpose of pursuing 'objective' truth" (cited in Flick 1992, 179).

Four basic types of triangulation were identified by Denzin and Lincoln (1994):

1. Data triangulation is the use of a variety of data sources in a study.
2. Investigator triangulation is the use of several different researchers or evaluators.

3. Theory triangulation is the use of multiple perspectives to interpret a single set of data.
4. Methodological triangulation is the use of multiple methods to study a single problem.

Janesick (1994, 214–15) suggested a fifth basic type: interdisciplinary triangulation. She refers to the problem of the dominance of the discourse by a single discipline and uses the example of psychology in education. She claims that, because of psychology's dominance, there is a danger of aggregating individuals into sets of numbers and that this has moved researchers away from an understanding of lived experience. She advocated incorporating other disciplines, such as art, sociology, history, dance, and architecture, into the discourse.

Hammersley and Atkinson (1995, 230) expanded the concept of data triangulation to include the comparison of data relating to the same phenomenon but deriving from different phases of the fieldwork and different points in the temporal cycles. They noted that data triangulation can be based on the accounts of different participants, including the ethnographer, located in different settings.

Hammersley and Atkinson further suggested that to maximize the results of triangulation, the observers should be as different as possible and should have different roles in the field (231). I will return to this point in the discussion on teamwork in chapter 5. Hammersley and Atkinson noted, however, that investigator triangulation does not usually occur, even when ethnographers do team research, since team members collect information on different aspects of a phenomenon. This is not true for RQI.

Sometimes the term *triangulation* is used to describe the combination or mixing of qualitative and quantitative research (Flick 1992, 177). This is not how the term is used in connection with RQI.

Hammersley and Atkinson (1995, 232) cautioned that simply aggregating more and more data does not necessarily produce a more complete picture. They suggested that having different types of data may be more important than the quantity of data.

While most social scientists are comfortable using the metaphor of triangulation, Richardson (1994) would like to see triangulation replaced with *crystallization*. According to her, triangulation assumes that there is a "fixed point" or "object" that can be triangulated. She rejects the notion that there can be a single, or triangulated, truth. She suggests that crystals are prisms that reflect the external and their refraction creates different colors and patterns. "What we see depends upon our angle of repose" (522). According to Richardson, crystallization provides a deepened, complex, but partial understanding of the topic. "Paradoxically, we know more and doubt what we know" (522).

Bogdan and Biklen (1998, 104) argued that when the goal of research is understanding rather than discovering laws, triangulation is not desirable and advised against the use of the term since it "confuses more than it clarifies, intimidates more than enlightens." They recommend describing the different data collection techniques used.

Willis (2007) was even more emphatic in rejecting the use of the term *triangulation*. Willis argued that triangulation is an extension of the idea of validity, a concept that is "not a core issue" when the assumption is that reality is socially constructed, and "thus there are multiple perspectives on reality." Willis rejected the need to try to eliminate all but one true reality. Finally he rejected triangulation as a general rule for research since a well-designed study using only one method, "such as interviewing," may be better than a study using multiple data collection methods. Willis (2007, 220–21) noted that there are alternatives to triangulation that can increase the credibility of the research including (a) **member checking**, defined as checking the emerging conclusions with the participants in the study (see ch. 4, "Checking Back with Informants"); (b) participator research, where the participants are actively involved in the formulation of conclusions (see ch. 9, "RQI and Participatory Research"); (c) peer review, defined as involving other scholars in the research; (d) researcher journaling; and (e) audit trails. It should be noted that RQI is based on involving other scholars in the research, and member checking, participator research, peer review, journaling, and use of audit trails are consistent with RQI.

While recognizing the need for caution about the use of the term *triangulation*, I believe as long as the focus is on ensuring different perspectives and there is no assumption that the goal of triangulation is finding the single truth or the one solution to an issue, triangulation can be a useful concept for improving the RQI process.

Triangulation does not assume there is a single truth or one solution to an issue.

Triangulation and RQI

The example of the interview with the director of the not-for-profit organization illustrates triangulation using a multidisciplinary team of investigators, combining direct observations with semistructured interviewing, and combining information collected in advance with the interview process. It also illustrates the power of teamwork for triangulation. The synergy resulting from ensuring multiple perspectives based on intensive teamwork is absolutely crucial for RQI.

When applied to RQI, the multidisciplinary team works together to collect data through semistructured interviews, through observations, and from information collected in advance of the RQI. Additional triangulation results from the different perceptions, theories, methods, and academic disciplines of the different team members (including the insiders on the team). The assumption is that for most situations there is no one best way to obtain information and that, even if there were, it could not be known in advance. Successful triangulation for RQI depends upon conscious selection of team members who can bring different perspectives, research techniques, theories, and disciplinary backgrounds. It then depends upon the team aggressively pursuing data from different sources. The issue of the selection of team members will be considered below in the sections "The Use of Teams" and "Diversity."

Intensive team interaction as an element of triangulation (along with team interaction for the iterative analysis and additional collection of data) permits the RQI team to significantly reduce the time in the field. The lone researcher may require a prolonged period to triangulate a single piece of information using only two or three different sources of information. Given the difficulty of both listening to others and collecting information on the physical context, while at the same time recording the information, the lone researcher may be forced to approach these sequentially. Without team interaction, a team of three researchers might be expected to require even longer (when the time of the individuals is added together) to collect the same information as the lone researcher. Working by themselves they still face the difficulty of focusing on several sources of data at the same time. There might be an increase in the triangulation with the greater number of observers, but even this is not certain, since without team interaction team members may not know the lines of inquiry or observations of the other team members.

Triangulation requires intensive team interaction.

When these same individuals interact with each other during the research time, tasks can be coordinated and shared. While one member leads the interview, another may assume responsibility for taking notes or recording the interview. At the same time, this team member may be observing for visual context and listening to the answers of the respondent from a different perspective based on different technical expertise. The communication between the team members, either directly or through their interaction with the respondent, provides immediate help and feedback to the other team member. The relationships that might take a lone researcher a significant amount of time to recognize may be revealed immediately and serve for new lines of inquiry. These in turn can result in additional cycles of triangulation. The possibility of triangulation based on different pieces of information and different interpretation of the same piece of information has suddenly multiplied many times. If team members are of different ages, gender, or ethnic identity, the potential for triangulation increases exponentially.

The presence of several team members makes it possible to collect more information per unit of time than can be collected by a lone researcher. The collection of additional information, however, is not the source of the strength resulting from the use of a team for RQI. The strength results from the team interaction and the impact of this interaction on the ability of the team to carry out data collection quickly but without being rushed.

Team interaction also has the potential for increasing the efficiency of data collection by allowing better decisions on what information is really needed. The observation of a team member that a certain line of inquiry may not be relevant has the potential of saving both data collection time and data analysis time. The assumption is that two heads are better than one, especially if the one is preoccupied with carrying on an interview or collecting some other information. The improved on-the-spot decision making concerning the information to be collected contributes to increased effectiveness of the data collection process.

Concepts and Illustrative Techniques Associated with the Use of Teams

Use of Teams

By definition, RQI cannot be done by one person. The expertise brought to the situation by the team members and their willingness to work closely together may be the most critical components of RQI. Team members are chosen based on the specific situation being investigated and the availability of different individuals to serve on the team. Team members should represent a range of disciplines that are most relevant to the topic. For example, an RQI team investigating health practices might include a social worker, a medical doctor, a traditional healer, and a public administration specialist. An agricultural development RQI team might include an agricultural economist and an agronomist. The mix of specific disciplines on the team often is not as critical as having different disciplines represented. It is important for team members to understand the rationale for a team effort and explicit discussion of what different disciplines are expected to contribute to the team effort can be useful. The selection of team members and getting the team to work together are two of the most important responsibilities of the team leader. These issues are discussed in the section on "Team Leadership."

> The disciplinary specialization of each team member usually is not as important as having different disciplines represented on the team.

Diversity

Since one of the keys to RQI is the use of a team composed of individuals who are looking for different things, listening for different answers, remembering and putting together different information, and relating to the respondents in different ways, diversity on the team is extremely valuable. It should not be assumed that, just because individuals have different academic disciplines, they will bring the diversity in perceptions, theories, and methods needed. If this type of information is not already known, potential team members should be asked about it while the team is being assembled. While the specific situation being investigated will influence the types of diversity that are most relevant, age, gender, and ethnic identity should be considered.

Insiders/Outsiders

Whenever possible, teams should be composed of a mix of insiders from and outsiders to the situation being investigated. Outsiders are able to share experience and knowledge from other situations and their participation can be extremely valuable to the insiders in identifying options and in noting constraints that might otherwise be overlooked. At the same time, outsiders gain insights and knowledge from insiders that can guide their understanding of other situations they might investigate in the future. See chapter 5, "Outsiders and Insiders in Between" for more information on the role of the insiders on the RQI team (also see ch. 10, "Insiders as Local Team Members"). Participation of insiders as full team members is one way of putting people first. Chambers (1991) noted that

> Where people and their wishes and priorities are not put first, projects that affect and involve them encounter problems. Experience shows that where people are consulted, where they participate freely, where their needs and priorities are given primacy in project identification, design, implementation, and monitoring, then economic and social performances are better and development is more sustainable. (515)

The RQI team should include a mix of insiders and outsiders.

Small versus Large Teams

RQI requires a minimum of two team members. Once the minimum requirements for disciplines and diversity have been reached, smaller teams of four or five members are preferred to larger teams. It has been my experience that members of large teams are more likely to engage in team interaction not related to triangulation, often just talking to one another, and are less likely to use team interaction to guide listening to and learning from others. Large teams often intimidate respondents, are more likely to be conservative and cautious, and take longer to produce a report and recommendations (Chambers 1983, 23).

Semistructured Interviews and Team Interviewing

Semistructured interviews by the RQI team provide numerous opportunities for developing understanding of situations as team members representing different disciplines initiate varied lines of inquiry and raise issues that otherwise could be overlooked. Two related issues concerning team interviewing should be noted. First, the individual or group being interviewed must not feel that a gang of tag-team interviewers is attacking them. A brief explanation for the presence of a team instead of an individual interviewer may be appropriate. Deliberate pacing of the interview, with increased efforts to maintain a conversational tone, may be needed. As discussed in chapter 5, the team needs to be sensitive to the fact that there are some situations where the interview cannot be done by a team, either because of the respondent's general discomfort with a group or because of the nature of the topic. In these very limited situations, one member of the team should do the interview. Second, team members must feel comfortable working together. This requires carefully listening to the questions by other team members and politely interrupting when appropriate. Conversely, team members must be comfortable being interrupted. A brief practice session with team interviewing may be useful (see appendix C on learning RQI). Careful reviews by the team following the first several actual team interviews will help everyone understand the specific behaviors that will be most beneficial. Prior agreement by the team on which team members will take the lead on specific topics is useful, but should not be interpreted as assigning topics exclusively to only one person. Conducting an interview requires the active participation of all team members. Intensive team interviewing is not based on sequential interviewing by members of the team, but on joint interviewing.

Team Observing

Direct observation is an important rapid qualitative research tool for reconsidering data collected in advance, providing multiple checks on data collected from interviews, and suggesting additional topics for interviews. Direct observation can prevent RQI from being misled by myth (Chambers 1980, 12). "Try it yourself" is an abbreviated form of participant observation in which team members undertake an activity themselves. Doing so allows insights and prompts the volunteering of information that otherwise might not be accessible (Chambers 1991, 524). Depending upon the situation, several specific direct observation techniques have been found useful. Where locally accepted, a camera can be an extremely important research tool. Photos can be used to document conditions before an intervention. Smartphones can be used both to take and share photos (see chapter 7 on use of technology). The preparation of sketch maps provides powerful visual tools that encourage the RQI team and local people to view community issues from a spatial perspective (see "Mapping," below). The use of proxy or unobtrusive indicators, such as the presence of drug paraphernalia, a sewing machine in a rural household, or changes in the construction materials for roofs, can provide insights about conditions and changes, especially when the local participants identify these indicators as relevant (see "Unobtrusive Observations," below).

Team Collection of Information in Advance of the RQI

Combining information collected in advance with information gathered from semistructured interviews and observations can significantly improve the methodological strength of RQI. Chambers (1980, 8) noted that despite the wealth of information in archives, including annual reports, reports of surveys, academic papers, and government statistics, rapid research teams often ignore these sources of data. The failure to collect basic data in advance of the RQI means that field research time is wasted in collecting already available data. Moreover, important research leads and topics suggested by previously collected material may be missed. The structure of the RQI process makes certain types of information collected in advance especially relevant. For example, minutes of past board meetings can be particularly relevant to an investigation of a not-for-profit organization, while maps and aerial photos may be relevant for a team looking at community activities.

A team effort at collecting information in advance uses the varying expertise of the different team members to identify information that will be most relevant. This increases the chances that the needed information will be available to the team (see ch. 10, "Materials Collected in Advance," and ch. 7, "Enhanced Teamwork").

Concepts and Illustrative Techniques Associated with Interviewing

Semistructured Interviews

Semistructured interviews based on guidelines are the key to RQI. The most important way of learning about local conditions is to ask local people what they know. The goal is to get people to talk on a subject and not just answer questions. Sufficient time must be invested to establish rapport and to explain the purpose of the RQI. As early as the 1930s, Webb and Webb (1932, as cited in Burgess 1982, 107) called this type of interviewing "conversation with a purpose."

Semistructured interviews are the key to RQI.

The interview should be a dialogue or process in which important information develops out of casual conversation. Metzler (1997) defined a creative interview as "a two-person conversational exchange of information on behalf of an unseen audience to produce a level of enlightenment neither participant could produce alone" (12). The key to successful informal interviewing is to be natural and relaxed while guiding the conversation to a fruitful end. "Talk with people and listen to their concerns and views" (Rhoades 1982, 17). Metzler suggested that "if you want candor—you want human responses rather than defensive exaggerations and false facades—try revealing a little of yourself in the conversation." He also suggested that striving for technical perfection can intrude on candor and that it is better to just have people talking (Metzler 1997, 2).

It is important to avoid the opinion poll syndrome, in which the RQI team drives up and jumps out with clipboards in hand, ready to interview. The RQI team needs to be especially sensitive to the fact that people may be suspicious of outsiders (Rhoades 1987, 119–20). The team needs to recognize that the time of the respondents is valuable and the team needs to be able to accommodate the respondents' schedules. Because the assumption is that an RQI is based on collaboration between the local people and the team, payment for interviews is usually inappropriate. However, compensation for lost income and actual costs such as transportation and expression of appreciation, including symbolic small gifts, may be very appropriate.

Even though the semistructured interview is flexible, it is controlled (Burgess 1982, 107). It has been suggested that the RQI must encourage respondents to relate experiences that are relevant to the problem, and discuss these experiences naturally and freely.

Experience has suggested that there are specific things that will improve an interview. Open the interview with a "Grand Tour" question (McCurdy et al. 2005, 38). Such a question might be "So tell me something about yourself" or "How did you happen to get here?" I knew I had found a good "Grand Tour" question when I asked one of the state farm managers in Poland to tell me about his farm. He replied by telling me that at the end of the last ice age, as the glaciers melted, they left different types of soils in different places on his farm. I expected to be there a long time with that opening! It turned out this was relevant information for the types of enterprises on this farm.

Wolcott (2005, 104) suggested becoming an active and creative listener in order to improve the interview process. This means more than being an attentive listener. It means using the listening to better play an interactive role and thereby make a more effective speaker out of the person talking. Wolcott also suggested talking less and listening more, making questions short, and planning interviews around a few big issues (105). Other ways of improving interviews include the use of culturally appropriate nonverbal behavior, such as paralanguage, voice, and eye contact (Metzler 1997, 52–53). The grunts and noises an interviewer makes in response to comments—the "umms," "uh-huhs," and "mmmmms"—are called paralanguage (52). Studies on the effects of these sounds leave no doubt that they contribute to rapport. People speak longer when they hear those kinds of responses. The tone of voice carries subtle but effective meaning. Experiments suggest that when words clash with the tone of voice and facial expression, people tend to believe the nonverbal aspects. Various studies suggest that eye contact enhances response and that people tend to look at the other person more while listening than while talking. However, it should be noted that in some cultures eye contact, especially between people who are not well known to each other, is considered rude.

One of the keys to successful interviewing is learning how to probe effectively, "that is, to stimulate an informant to produce more information, without injecting yourself so much into the interaction that you only get a reflection of yourself in the data" (Bernard 2011, 161). Sometimes it is necessary to repeat a question, which should be done without expanding or elaborating. When you expand or elaborate you are likely to suggest an answer or to change the question (Wolcott 2005, 105). Depending on the culture, one of the most effective and difficult-to-learn probes is the silent probe. It consists of just remaining quiet and waiting for an informant to continue. Ely (1991) noted that it may be especially difficult to remain silent, and thus let the respondent continue, when what is said strikes a chord with the interviewer and he or she feels compelled to share because, after all, "we've been there." (60). Interviewers are reminded to distinguish between a pregnant pause and dead silence (Wolcott 2005, 105). Another kind of probe consists of simply repeating the last thing an informant has said and asking them to continue. This probe may be especially useful when an informant is describing a process, or an event (105).

Keeping the interview moving naturally requires a few comments and remarks, together with an occasional question designed to keep the subject on the main theme, to secure more details, and to stimulate the conversation when it lags. Examples of nondirective probes include

Give me a description of . . .
Tell me what goes on when you . . .
Describe what it's like to . . .
Tell me about . . . (or) Tell me more about that.
Let's see [pause]. I'm having trouble figuring out how I can word this.
Give me an example.
How might someone do that?
How important is that concern?
So, the message you want me to get from that story is . . .
Say more.

Keep talking.
Don't stop.
How come your energy level just went down?
What am I not asking? (adopted from Silverman 2011)

RQI team members need to have understanding and sympathy for the informant's point of view. "They need to follow their informants' responses and to listen to them carefully in order that a decision can be made concerning the direction in which to take the interview. In short, researchers have to be able to share the culture of their informants" (Burgess 1982, 108).

There are techniques for "jogging" respondents' memories that will improve the accuracy of their responses. When time has passed since the events being discussed, respondents can be asked to begin by recalling personal landmarks. Once the list of personal landmark events is established, it may be easier to recall other events in relation to them (Bernard 2011, 185). While there are several specific techniques that can be used to help deal with memory errors, the one I believe is most relevant to RQI is asking respondents to consult records, such as bank statements, telephone bills, and minutes of meetings. Often, copies of these documents can be requested. There are smartphone apps that allow the team to make copies of documents on the spot. (See chapter 7 on technology.)

Direct quotes are critical for telling the story of the local participants. In RQI reports by the team, direct quotes can be used to establish authority and authenticity. Direct quotes employing idiosyncratic speech, figures of speech, metaphors, and local sayings add human color. Even when these quotes are in English, they may need to be translated. The really good quotes that meet these needs may not occur spontaneously. Journalists have developed techniques that can facilitate this process and are relevant to ethnographic interviews.

Metzler (1997) suggested several techniques for encouraging people to talk more "quotably" (102). He suggests asking for a simile or an analogy. The question is "What is it like to . . . ?" The inclusion of metaphors in questions or comments will often elicit responses that include metaphors. However, if the respondent uses your metaphor in the response, it is not authentically his or hers, and another probe may be necessary. Other ways of getting rich responses are to use questions like:

Is there an old Texas (substitute local place) saying that covers this situation?
Can you put a label on this for me?
How would you explain this to your nine-year-old daughter?
The situation you describe sounds like a sophisticated Monopoly game—are you winning or losing? (104–5)

Sometimes respondents or informants will telegraph a quote. There is a need to be especially attentive when you hear phrases like "In my view" and "Long years of experience have prompted me to suggest" (105).

Metzler (1997, 106) suggested that dramatically showing your appreciation for a quotable remark can encourage the respondent to produce more, since they now know you are looking for them. He also suggested that the interviewer needs to stay particularly alert for personal asides. These can differ significantly in content and style from the more structured responses provided in response to semistructured questions.

As a general rule, interviews should be conducted under conditions most relevant to and revealing about the local system being investigated. For example, an RQI on health care should include interviews in the clinics where services are provided, while an RQI on agriculture should include interviews in farmers' fields, where the team can witness farmers' behavior. If the interviews with the farmers in the western Sudanese village had not occurred in their fields, it is unlikely that the role trees that harbor birds play in decisions on crop rotation would have ever surfaced. Actual observation permits the identification of new topics for discussion. Conducting as many interviews as possible at the site of the action being investigated is an important part of observation (see "Participant Observation" below). The RQI team should always note where interviews were conducted.

The RQI team needs to be aware that the answers they get can be influenced by numerous factors. In many cases, there is little that the team can do about these, other than to be aware that they exist and to be cautious about the answers. Different people asking the same question will sometimes get different answers, and even the same person asking the same question but in different settings may get different answers. Factors that influence responses include race, sex, age, and accent of both the interviewer and the respondent; the source of funding for a project; the level of experience of the interviewer; cultural norms about talking to strangers; and whether the question is controversial or not (Bernard 2011, 177). In many cultures, women get different responses to questions than men.

Sometimes informants will tell you what they think you want to hear, in order not to offend you. The way questions are worded can make a difference in the answers you get, especially for more threatening questions (Bernard 2011, 181). It is possible to get the results you expect not simply because you have correctly anticipated things but because you have helped to shape the responses merely by the RQI team's presence. For example, the presence of an RQI team listening carefully to the concerns of local people can impact their views as to whether anyone is interested in their problems and thus whether they should try to influence things.

Use of Illustrative Topics and Short Guidelines

The disagreement among ethnographers over the extent to which they should develop hypotheses and detailed guidelines before starting their work extends to RQI. Some RQI teams begin with pages of detailed guidelines that are to be followed closely, with all questions being asked. At the other extreme, some RQI teams begin with only a brief list of topics. Failure to offer specific questions appears to be premised on the belief that interviews should be very general and wide ranging, especially since the team is exploring and searching for an unknown number of elements. It is claimed that a framework prepared before beginning an RQI can predispose team members toward their own ideas, thereby blocking opportunities to gain new insights.

My experience suggests that short guidelines prepared in advance can be useful as long as they are not relied on too much. "In this early phase, the researcher is like an explorer, making a rapid survey of the horizon before plunging into the thickets from which the wider view is no longer possible" (Rhoades 1982, 5). While one may begin with guidelines, important questions and the direction of the study emerge as information is collected. Interviews should be planned around a few big issues. The interview then focuses on exploring the dimensions of an issue instead of details from a long list of questions (Wolcott 2005, 105). Bernard (2011, 163) suggested you can get longer responses by making your questions longer. Instead of asking sponge divers, "What is it like to make a dive into very deep water?"

ask "Tell me about diving into really deep water. What do you do to get ready and how do you descend and ascend? What's it like down there?" While Bernard noted that asking longer questions does not necessarily produce better responses, they are likely to keep informants talking and "the more you can keep an informant talking, the more you can express interest in what they are saying and the more you build rapport" (217).

Calling the guidelines "illustrative interview topics" instead of "interview questions" provides a reminder that they are not a list of questions to be asked. Guidelines should not be viewed as an agenda to be diligently worked through, but as an aid to memory and a reminder of what might be missed (Bottrall 1981, 248). "Not everything needs to be known. The key . . . is to move quickly and surely to the main problems, opportunities, and actions" (Chambers 1983, 25).

> The goal of RQI is to have people tell their stories and not to have them answer the RQI team's questions.

Use of an Audio Recorder

I appreciate Patton's (2002, 380) often quoted statement: "As a good hammer is essential to fine carpentry, a good tape recorder is indispensable to fine fieldwork." Bernard (1995, 222) cautioned against relying on your memory in interviewing and recommends using an audio recorder in all cases except where respondents specifically ask you not to. He noted that, even when there are no plans to transcribe interviews, recordings can be used to fill in missing information from notes taken during the interview (223). Chapter 7 provides extensive information on choices concerning the use of audio recorders.

Experience teaches that equipment will fail, so the RQI team needs to take extra precautions, have backups, and be prepared for the backups also to fail (see ch. 10, "Interview Notes in Addition to Recording").

> Finally, never substitute recording for note taking. Take notes during the interview about the interview. . . . What were the physical surroundings like? How much probing did you have to do? Take notes on the contents of the interview, even though you get every word on the machine. (Bernard 2011, 171)

Never substitute audio recording for note taking.

Use of Interpreters

Many qualitative researchers subscribe to the hypothesis that human beings speaking different languages live in different worlds, with language acting as filters on reality and molding perceptions of the worlds (Werner and Campbell 1970, 398). Ideally, all members of an RQI team should speak the language of the people from the situation being investigated. In practice, however, one or more members of a team may not speak the local language and an interpreter must be used. There is no excuse for failing to learn and use appropriate greetings. Knowledge of numbers and even a very few keywords can allow a team member to appear to understand more than he or she actually does. This can improve rapport and the quality of the translation. Interpreters should be chosen carefully to ensure that they understand technical words that are likely to be used in the questions or answers. Werner and Campbell (1970) suggested that interpreters be chosen on the basis of their competence in the target language rather than in English. Before the interview, the team should go over the interview strategy with the interpreter, emphasizing that the team is interested in more than just "answers" to "questions." Werner and Campbell suggested providing the interpreter with a reformulated English version of the interview guidelines. They identified the following steps for the preparation of this reformulated version:

1. Get a good dictionary and a good thesaurus of English.
2. Rewrite all complex sentences as simple sentences.
3. Eliminate all metaphors and idioms.
4. Look up all the keywords in the simple sentences and include two sets of possible paraphrases for each sentence: (a) paraphrases which are acceptable and (b) paraphrases which are not. The researcher presents the interpreter with both sets and explains that translations should be close to the acceptable paraphrases (408–9).

The interpreter should not be physically between the speaker and the person being interviewed, but rather beside or slightly behind the person or persons leading the discussion so that her or his function is clearly indicated. The team member should speak in brief sentences using a minimum number of words to express complete thoughts. The interpreter should be given time to translate before proceeding to the next thought. The team member should talk directly to the respondent, as if the respondent can understand everything said (Bostain 1970, 1). It is especially important to record interviews where an interpreter has been used, since this permits the translation to be checked.

Selection of Respondents

In most situations people can be interviewed about their own experience or their knowledge of the broader system beyond their own experience. Differentiating between these types of interviews can result in better information. Persons interviewed about their experience are referred to as **individual respondents**. They should be selected to represent variability. It should be clear to both the respondent and team members asking the questions that the questions concern only the individual's knowledge and behavior, and not what he or she thinks about the knowledge and behavior of others. Persons interviewed about the broader system are often referred to as **key informants** and should be selected because of their experience and knowledge. Unless the terms *informants* or *respondents* are modified with the words *individual* or *key*, they can be used interchangeably. *Participants* is another term that is used interchangeably with *respondents* and *informants* to identify the persons interviewed as part of the RQI process. The term *subjects* is generally not used.

Interviews should be conducted with an opportunity sample of purposely selected individual respondents. These individual respondents should be chosen because they represent a wide range of individuals in the situation being investigated and those chosen should not be limited to what is assumed to be representative or average. The research team for the Polish state farm study explicitly requested interviews with successful managers, unsuccessful managers, midlevel managers, a tractor driver, fieldworkers, Ministry of Agriculture personnel, bankers, cooperative officials, and nonstate farmers. The RQI team needs to be sensitive to the bias that would be introduced if only one gender was interviewed when both are involved in a situation.

> Seek out the troublemakers.

Following Honadle's (1979, 45) strategy for avoiding biases when investigating organizations, the RQI team could ask for the names of one or more individual respondents who are known to disagree with all decisions, generally promote trouble, and never cooperate with programs. Responses from these persons can provide valuable cross-checks and insights not available from other interviews.

Key informants are expected to be able to answer questions about the knowledge and behavior of others and especially about the operations of the broader environment. They are willing to talk and are assumed to have in-depth knowledge. Key informants for a study of a school system might include student leaders, administrators, school board members, and leaders of parent-teacher associations. It is usually worthwhile to ask which people are most knowledgeable and then seek them out. Key informants may need to be interviewed several times with information from one session or information from other respondents identifying topics for the next session. The time constraints of an RQI should not be used as an excuse for not interviewing the same person several times if this is appropriate (see ch. 10, "Follow-Up Interviews with the Same Person").

Focus Group Interviews

Any interview by an RQI team of a group of respondents can be considered a focus group interview since focus group interviews can be very informal or highly structured. Focus group interviews can be extremely useful in collecting certain types of information. In a focus group interview one or more RQI team members moderate an interview with a group of respondents. The hallmark of focus groups is their explicit use of group interaction among respondents to produce data and insights that would be less accessible without the interaction found in a group. The RQI team needs to be clear that interaction among respondents is encouraged.

Morgan (1997, 2) defined focus groups as a research technique that collects data through group interaction on a topic determined by the researcher. In essence, it is the researcher's interest that provides the focus, whereas the data themselves come from the group interaction (6). According to Krueger and Casey (2009, 6), a focus group consists of (a) people, who (b) possess certain characteristics, (c) provide qualitative data, (d) in a focused discussion (e) to help understand a topic of interest. Krueger and Casey identified the need for participants to feel comfortable, respected, and free to give their opinion without being judged as conditions necessary for focus groups to work (9).

Group interviews can be used in some cultures to collect information on topics where an individual may be penalized if he or she replies truthfully, but where a group talking about the community may not feel threatened (Chambers 1980, 14). Wellner (2003 as cited in Stewart, Shamdasani, and Rook 2007, 4) noted that focus groups could be used in natural settings with real social groups, the type of conditions where focus groups are most likely to be used during an RQI. Often, similar topics can be taken up in interviews with "key informants." Focus group interviews where individuals are free to correct each other and discuss issues can identify variability within the community and prevent an atypical situation from being confused with the typical.

My experience suggests that group interviews may reveal what people believe are preferred patterns as opposed to what actually exists. A very detailed description of the local crop rotation system by a group of farmers in the village in western Sudan was later found not to be practiced by any of them exactly as described (Beebe 1982). Even when some topics have been covered by a group interview, the same topics should still be covered with individuals. The question changes from "What do local people generally do?" to "What do you do?"

The presence of others often influences answers, and who is present during an interview may need to be noted. The presence of authority figures can almost always be expected to influence comments. For the western Sudan village study, about the only time the team was alone with individuals was during visits to their fields. The presence of neighbors and especially the village leader seemed to increase the number of positive comments about the government. The initial report that useful agriculture extension agents had visited the village was not confirmed in the interviews with individuals. One participant remembered only one visit, some twenty-five years in the past, during which the extension agent had brought tree seeds that did not survive. In another individual interview, the participant suggested that the agriculture extension agents who had visited the village were useless and contrasted them to the technicians from the water department who had helped with a well in a nearby village.

Other Concepts and Illustrative Techniques

Comparing and Sorting Objects

Physical objects can be used as stimuli for eliciting responses. Objects can be three-by-five-inch index cards with names of people, places, or objects; photographs; or the actual objects, such as plants or birth control devices. In a triad test, three objects are presented and the respondent is asked to either identify the one that does not fit, choose the two that seem to go together, or choose the two that are the same. Discussion often centers on what the respondent thinks it means to be the "same" (Bernard 2011, 233). In a pile sort, the respondent is asked to sort objects (often cards with the names of things or concepts written on them, although they could be photographs or physical objects) into piles, putting things that are similar together in a pile. Discussion can center on what is meant by "similar." If the respondent asks whether something can be put in more than one pile, Bernard (2011, 233) suggested three possible responses: saying "no" because there is one card per item and a card can only be in one pile at a time, making up a duplicate card on the spot, or asking the respondent to do multiple pile sorts of the same objects. These techniques have been extensively used for "cultural domain analysis." When these techniques are used as part of an RQI, the objective is not to analyze cultural domain but to stimulate general discussion. Given the time constraints of RQI, the sophisticated analysis associated with cultural domain analysis would not be appropriate. In an RQI, these techniques are as likely to be administered to a group as to individuals.

Participant Observation

Pelto and Pelto (1978) noted that "every individual is a participant observer—if not of other cultures, then at least of one's own" (169). Participant observation as an essential qualitative research technique, however, requires much more than simply being there and passively watching. It requires systematically exploring relationships among different events and recording what is seen and heard. Ely (1991) spoke about intensive observing, listening, and speaking. Social scientists generally recognize that the role of the participant observer can range from full participant, defined as actually living and working in the field as a member of the group over an extended time, to mute observer. The essential requirements for participant observation are that people must feel comfortable with your presence and allow you to get close enough to observe (Bernard 2011, 256).

> Interviews should be conducted where listening can be combined with observing.

In the section on semistructured interviewing (see "Semistructured Interviews," above), I emphasized that interviews need to be conducted in a relevant setting where listening can be combined with observing. I strongly agree with Ely that "interviewing cannot be divorced from looking, interacting, and attending to more than the actual interview words" (1991, 43).

Even though there is not a lot of time during an RQI to do participant observation, there are more opportunities than one might expect. Every interview is an opportunity for participant observation and the results of every interview should include observations. There are sometimes opportunities to do things that are observed. My attempt to spread some chemical fertilizer in a rice field in the Philippines was not very successful, but the comic relief it provided to the farmers resulted in some wonderful dialogue on how one learns new farming practices. The results of a "try it yourself" can significantly enhance the results of an interview. Finally, even during an RQI there are opportunities that should not be missed for sharing time outside the boundaries of interview sessions. Whenever appropriate and possible, the RQI team should spend the night in the local setting. Critical events during the early mornings, evenings, or nights may not arise in interviews during office hours. Sharing of meals provides opportunities for informal discussions and follow-up to topics discussed during formal interviews. If it is not possible to share meals, it is almost always possible to share tea or coffee. The RQI team needs to be especially sensitive about imposing on the hospitality of others and must be ready to return favors.

Unobtrusive Observations

Unobtrusive observations are direct observations done inconspicuously in which the participant may or may not be aware of the observation when it occurs. Often, physical objects are observed and used as proxy indicators for social behavior. "A well-worn path provides excellent, though incomplete, evidence concerning the volume of traffic between two communities; the different states of disrepair of buildings provide possible indexes of relative affluence; the extensiveness of refuse heaps give testimony to duration of occupation" (Pelto and Pelto 1978, 115). Other examples of unobtrusive observation include graffiti in public toilets, the age of automobiles in junkyards, and the presence of television satellite dishes or antennae in rural neighborhoods. The strength of these observations for an RQI is that they open new areas for discussion and for getting at the interpretations that local people provide of them. Unobtrusive observations are appropriate in public places where questions of privacy do not exist. Usually such observation is considered as having insignificant potential to harm participants. Fluehr-Lobban (1998, 184) noted that even if unobtrusive observation does not harm participants, this does not exempt researchers from providing full disclosure (see "Team Observing," above).

Folktales and Other Verbal Lore

Folktales, myths, songs, proverbs, riddles, and jokes can provide insights into local situations and interesting focal points around which to initiate discussion. Since one of the objectives shared by ethnography and RQI is to get people to tell their stories, the possibility of asking people to literally tell stories should not be overlooked. Even a rough content analysis of verbal lore can suggest possible themes for subsequent examination. Variations in the same story as told by different individuals can provide evidence of the diffusion of information (Pelto and Pelto 1978, 113).

Systems, Soft Systems, and Rich Pictures

A systems approach is implicit in RQI. Systems methodology provides an expanded set of conceptual tools and specific techniques for understanding how local people view their situation. It should be noted that many of these techniques are used by anthropologists and other social scientists but without reference to "systems." A system can be defined as a set of activities linked together for a purpose (Checkland and Poulter 2007, 9–10). For the purposes of RQI, it is useful to expand the meaning of a systems approach to include that the elements in the system behave in a way that an observer has chosen to view as coordinated to accomplish one or more purposes (Wilson and Morren 1990, 70). A systems approach initially considers all aspects of a local situation, but quickly moves toward the definition of a model that focuses on only the most important elements and their relationships to each other. Systems are always complex, and it is not possible to deal with all aspects of a system at the same time. The first task of an RQI team is to make a rough approximation of the system and to identify the elements that are most important for the specific situation being examined. It is very important to note that the elements in a system usually cannot be identified in advance, nor can decisions be made in advance as to which elements of a system are most important for understanding a given situation.

Soft Systems Methodology is an approach to systems thinking that is especially relevant to RQI. Soft Systems Methodology is associated with Checkland, who has identified several specific steps:

1. identifying a situation that has provoked concern
2. selecting some relevant human activity system
3. making a model of the activity
4. using the model to question the real-world situation
5. using the debate initiated by the comparison to define action that would improve the original problem situation (Checkland and Scholes 1990).

Checkland and Poulter (2007, 13) expanded on the use of the systems model to question the real-world situation by noting that the aim of soft systems is to find "changes which are both arguably desirable and also culturally feasible in this particular situation." Identifying changes that are desirable and culturally feasible are goals shared by RQI.

Umans (1997, 11–13) explicitly used Soft Systems Methodology for program planning in a Rapid Appraisal of the traditional health knowledge system of the Guarani Indians in Bolivia. After the soft systems model of the role of traditional midwives was compared with the real-world situation, it was decided that independent supervision of midwives was not being done. After a discussion of this discrepancy, it was decided to include this in a list of proposals to improve the existing situation.

A critical element of Soft Systems Methodology is to develop a "**rich picture**" of the situation under investigation. According to Checkland and Poulter (2007, 25) the rationale for making a rich picture is "to capture, informally, the main entities, structures, and viewpoints in the situation, the processes going on, the current recognized issues, and any potential ones." Drawing diagrams and pictures allows both individuals and groups to express and check information in ways that are often more valid than talk. Checkland and Scholes (1990, 45) argued the reason for this "is that human affairs reveal a rich moving pageant of relationships, and pictures are a better means for recording relationships and connections." The rich picture is literally a drawing to which all the participants in the activity are encouraged to contribute. Figure 3.1 is an example of a rich picture from the not-for-profit RQI.

Mapping

Mapping is the use of simple graphics, including drawings, pictures, and sketches. Mapping can be used for collecting data, presenting data, understanding data, and planning action. As a data collection technique, maps are drawn by local participants or the RQI team and can be used to illustrate important information about individuals, social groups, and the wider environment. Spatial mapping includes geographic and social composition of a community, building, or place. Spatial maps can include areas of activity, boundaries, key people, behaviors associated with location, and contextual factors such as income levels or ethnic groups. Spatial maps are only one kind of map, and mapping can be applied to almost anything. In addition to maps produced by respondents, Google maps can be used as a beginning point for maps that local participants are asked to correct or add to hard copies. In 2013 Google announced "Google Maps Engine Lite (Beta)" (https://mapsengine.google.com/map/splash?app=mp) for the creation of custom maps that allow for the import of spreadsheets of location and for styling and drawing options.

Other types of maps include network maps showing relationships between different people and groups, body maps showing the perceived

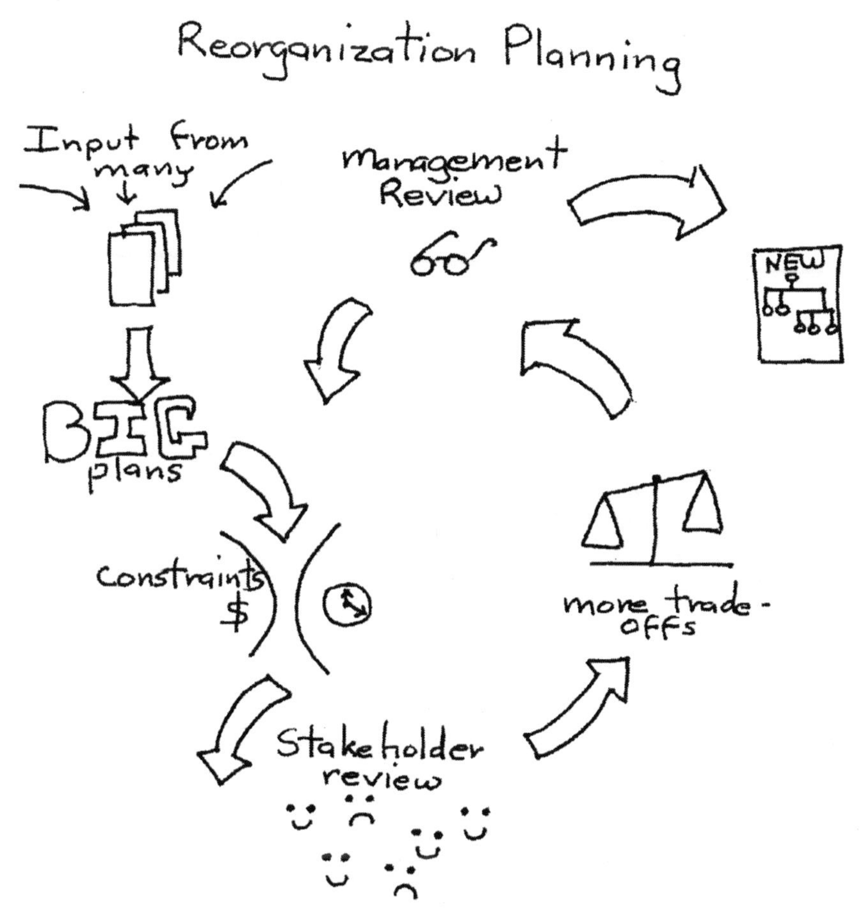

Figure 3.1. Example of a Rich Picture from an RQI

effects of substance use on an individual's health, and information maps showing the distribution of health or social conditions over time.

Mapping can quickly collect and present complex information in a simpler form. Other advantages of mapping are that it can be conducted with people regardless of age, literacy, or familiarity with social science research. Mapping can facilitate shared understanding between the RQI team and the community and it can identify areas in the community where interventions could be located. Mapping may provide for the participation of individuals who would not be comfortable speaking. Maps can identify relationships that otherwise might not be noticed and highlight nuances that might otherwise remain hidden. Figure 3.2 is from a Rapid Appraisal of the health system of the Guarani Indians in Bolivia and maps the relationships between actors and knowledge systems. Participants first identified relevant actors in the health system (listed in the circles) and then were asked to group together actors with

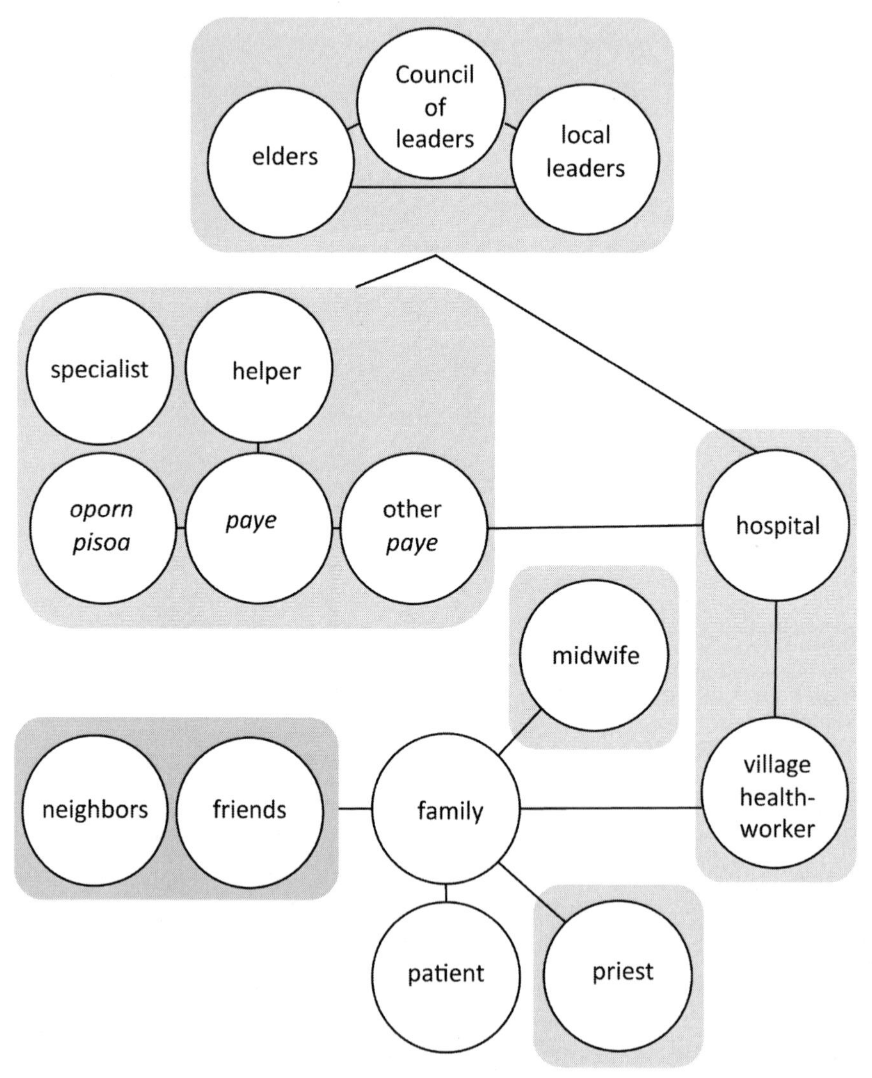

Figure 3.2. Map of Relevant Actors and Knowledge Systems (Guarani Indian health system, Bolivia)

good contacts (shown by the gray shading) and to map the relationships between individual actors, groups of actors, and sectors (lines). Lines may extend directly from one actor to another, from one actor to a grouping or sector, or from one grouping to another—all of which indicate distinctions in the various relationships (Umans 1997, 12). (See "Team Observing," above.)

Managing Data—Field Notes and Logs

Entire books have been written on **field notes** and there is very little agreement on what constitutes field notes and what you do with them once they have been collected. What is outlined here has worked for some of the teams on which I have participated, but always with some modifications to fit the specific situation. Because more than one person is involved in collecting and processing field notes for an RQI, it is extremely important that all parties agree on the format at the beginning.

Following the suggestion of Ely (1991, 69), I believe the term *field notes* should be reserved for the usually handwritten notes that are done as the data is being collected. The transcripts of interviews that were recorded can be called "transcripts." The term **log** is reserved for the repository of information from the field notes and transcripts in a format ready for analysis. Obvious, but sometimes overlooked, is the need to ensure that day of the week, date, time, location, and the name of the individual taking the notes are recorded as part of the field notes. The assumption has to be that loose pages will at some point be separated and thus all pages should be numbered.

Field notes do not have to be neat or even readable by anyone other than the person who takes them. They must however be written in a way that clearly indicates the difference between observations and reflections on the observations. Observations include what is heard, for example before an interview, and what is seen. Field notes based on careful, detailed observations as opposed to vague summaries can often help the observer avoid imputing false meaning to people's actions. Pelto and Pelto (1978) provided hypothetical field note entries that illustrate the difference between vague notes and notes that preserve the details.

> Vague notes: A showed hostility toward B.
>
> Concrete notes: A scowled and spoke harshly to B, saying a number of negative things, including, "Get the hell out of here Mr. B." He then shook his fist in B's face and walked out of the room. (70)

I have found it useful to take notes on interviews even when I have been responsible for leading the interview and the plans are to transcribe the recording of the interview. Note taking forces pauses in the process and can keep the interview from being rushed. Metzler suggested controlling the interview to accommodate note taking. Answers to questions can be repeated and you can ask if you have heard them correctly. If you have, the next statement can be something like "Okay, let me get this into my notes before we go on" (Metzler 1997, 116). Often, I have also used these pauses to consider where the interview was going and what should come next.

Even when interviews are recorded, there is a need for notes. The most common objection to taking notes during an interview is that it is too difficult for the interviewer to think about guiding the conversation and at the same time worry about taking notes. The use of a team during the RQI prevents note taking by the interviewer from being a problem, since someone other than the person leading the discussion can take the notes. Since all team members are expected to participate in the interview process, this can become complicated, and at times it will be useful to have several team members taking notes on different parts of the interview. Notes during a recorded interview serve two important purposes. First, they are needed for recording observations that are not audible. These might include a description of the settings, who else was present (or close enough to hear), the overall demeanor of the respondent, and nonverbal communications like a smile, a yawn, or the tapping of a finger on the desk. A second use of the field notes during a recorded interview is to ensure that important words are not missed because the recorder does not pick them up or that the entire interview is not missed because of a general failure of the audio recorder.

In addition to the observation by RQI team members, the field notes (and the log), will include reflections. These may be referred to as "observer comments," "analytic memos," or "researcher memos." They may include comments about whether a specific research technique was appropriate. These are conversations with oneself about what has occurred, what has been learned, insights, and leads for future action (Ely 1991, 80). Reflections, defined as any comments that are not observations, can be called **MEMOS**. I am spelling MEMOS with all uppercase letters to remind you of the importance of clearly identifying them in both the field notes and the log as your reflections, and not confusing them with observations (things heard or observed). MEMOS are sometimes recorded on the same tape as an interview immediately following the interview. MEMOS can be clearly identified, in both field notes and the log, by including them in brackets (or underlining them) and starting each with the word MEMO. A log might include MEMOS written at the time that the log was prepared and MEMOS added to the log at a later date. Each time a member of the RQI team adds a MEMO to the log, she should identify herself and give the date. For field notes, it is not possible to overemphasize the need to separate observations from reflections and to identify the reflections in the MEMOS.

> Anything in the field notes or the log that is not a direct observation should be identified as a MEMO, enclosed within brackets, and dated, and its author should be identified.

The log is the data upon which the analysis is conducted. The log is based on the field notes and is usually prepared at a location away from where the observations were collected. Missing words in field notes for interviews that were not recorded are filled in when preparing the log. The log will include the transcripts of recorded interviews. Just as in the field notes, it is critical to distinguish between what was seen and heard and reflections on the observations. Reflections should be clearly identified as MEMOS. MEMOS should be dated and for an RQI their author should be identified. Logs are a chronological record of what the team members learn and their insights.

Logs are usually most useful if typed, double-spaced, with very wide margins on both sides. Lines may be numbered. In the next chapter we will discuss adding codes, using words and not numbers, to the left margins and margin notes to the right margin (see ch. 4, "Data Condensation, Coding, and Margin Remarks"). Figure 4.2, in the next chapter, is an example of a log that includes a MEMO transferred from the field notes, codes, and marginal remarks.

Logs should be prepared within twenty-four hours of the experience. This may be equally important for the lone researcher and the RQI team, since forgetting begins as soon as the experience ends. The completion of the log as soon as possible is absolutely critical for the RQI team, since it is the basis for iterative analysis and additional data collection (see ch. 10, "The Twenty-Four-Hour Rule"). There are no hard-and-fast rules for RQI, but if there were, the completion of logs before moving on to the next data collection cycle would be one of them. Logs should contain as much detail as possible.

> The Twenty-Four-Hour Rule: Logs should be prepared within twenty-four hours of the interview.

Specific Strategies for Triangulation

Since the objective for triangulation of multiple sources of data is not to find a single truth but rather to expand understanding by seeking nuances and complexity, the team needs to be intentional about triangulating the data collected from the multiple sources. While collecting information by interviews or observation, it is the responsibility of the members of the team who are not leading the data collecting activity at the moment to ask, "What does this new information add to the existing understanding?" and, where appropriate, interject questions and comments. When the team begins the group review following a round of data collection and identifies possible conclusions, two questions need to be asked: (1) "What other data has been collected that is related to this issue and how does the other information impact understanding of this issue?" and (2) "What additional information still needs to be collected to better understand this issue?" Even when a topic has been explored in detail by a team member, another team member with a different background may be able to advance understanding based on his background. Since semistructured interviews are likely to be the primary data, the team needs to explicitly consider how observations and data collected in advance enhance understanding of issues provided by interviews.

> Triangulation of multiple data sources requires intentionality.

More than twenty concepts and techniques associated with qualitative research and RQI have been introduced in this chapter. If you are new to qualitative research, it may seem a bit overwhelming. To begin experimenting with RQI, all you need to remember is that your goal is to talk with people and get them to tell their stories, as opposed to answering your questions. The different techniques introduced in this chapter can be thought of as tools that can help you better do this. If you at least know a tool exists, you will be able to find it when you need it.

The theme of this chapter has been that two sets of eyes and ears are better than one. The techniques suggested should help make the best use of the extra eyes and ears as part of intensive teamwork. The theme of the next chapter is that two heads are better than one in figuring out what has been seen and heard and what should be seen and heard next before trying once again to make sense out of the data collected. The intensive teamwork implementing the iterative process of data analysis and data collection should help make the best use of the additional heads.

Additional Readings

For additional readings for this chapter, please see chapter 2; chapters 2 and 3 share a common list.

CHAPTER FOUR
ITERATIVE ANALYSIS AND ADDITIONAL DATA COLLECTION

Main Points

1. An iterative process is defined as a process in which replications of a cycle produce better and better results. For RQI an iterative process is a reflexive process and not a repetitive mechanical task.
2. The constant shifting between data analysis and additional data collection is an iterative or recursive process.
3. RQI is divided between scheduled blocks of time used for collecting information and blocks of time during which the team engages in data analysis and considers the next round of data collection.
4. The iterative nature of RQI allows for the discovery of the unexpected.
5. The importance of the intensive team interaction before each new cycle of data collection cannot be overemphasized.
6. Before the conclusions are final, the RQI team needs to share them with the people who have provided the information.
7. The joint preparation by the entire RQI team of the RQI report continues the intensive team interaction.
8. The Miles et al. (2014) model of analysis involves three steps: (1) condensing the data including coding and adding comments, (2) displaying the data, and (3) drawing conclusions.
9. The critical first step in the analysis process is dividing the log into thought units and applying codes to these units.

10. There are numerous tactics for generating understanding, including (1) identifying patterns and themes, (2) seeing plausibility, (3) clustering, (4) metaphor making, (5) counting, and (6) making contrasts and comparisons.

Iterative Analysis

An Example of Iterative Analysis

Beginning with the initial meeting with all of the directors of state farms in the Koshlyn region of Poland, the team heard shrill complaints about the high interest rates on loans. Initially, the RQI/RAP team focused on this as one of the new uncertainties that state farm directors faced. During one of the evening team reviews the decision was made to ask about the manager's understanding of the relationship between interest rates and inflation the next time a manager complained about high interest rates. This produced a vague explanation from the next manager interviewed about the real costs of borrowed money. This issue was placed on a back burner. A visit to a farm where tractors were lined up in the parking lot and crops were beyond ready for harvesting led to a discussion of why the tractors were not in the fields harvesting the crops. The director of the farm indicated credit was not available to purchase fuel for the tractors because the "high interest rates" on a previous loan had made it impossible to repay. Discussion by the RAP/RQI team on these results led to the decision to initiate a new line of questions with the next director on the traditional sources of cash needed for different operations. This led to an explanation of the old system, where state farms had easy access to very cheap loans from the state bank. These loans, however, were tools the Communist Party and the state used to limit the initiative and behavior of the state farms. For example, an initiative by a state farm to expand into a new crop, a new process activity, or even a new enterprise, like tourism, required a loan, and with the loan came close state control. Cash-flow management and timing of events to produce funds needed for the next activity had not been an issue with the old system and the less successful managers had failed to recognize the change in the system. The RAP/RQI team pursued this with the next director and discovered similar descriptions. The need for improvements in cash-flow management became one of the least expected but most important findings of the RAP/RQI team.

Iterative Processes and Qualitative Research

The example above illustrates the iterative process of beginning to tease out the essential meaning, then using this initial analysis to guide additional data collection, and then repeating the process. The almost constant shifting between data analysis and additional data collection is an iterative, or recursive, process. An iterative process is a process in which replications of a cycle produce results that approximate the desired result more and more closely. Each cycle of data analysis and data collection is expected to produce better and better results. This same process is labeled by some social scientists as a recursive process. A recursive process is a process that can repeat itself indefinitely or until a specified condition is met. One of the specific conditions that usually should be achieved during qualitative research is for the data to begin to repeat themselves. When this occurs, the additional data will not have a significant impact on the analysis. The goal of the analysis process is to draw conclusions from the data that can be shared with others in an economical and interesting fashion.

Iterative Processes and RQI

The RQI team, like other qualitative researchers, often begins with information collected in advance and then progressively learns from information provided by semistructured interviews and direct observations. For the RQI team, the data collection process, with its focus on getting the insider's perspective and the use of multiple sources, is enhanced by intensive team interaction. The RQI team, again like most individual qualitative researchers, engages in a process of iterative data analysis and collection of additional information. For the RQI team, there is a strict commitment from the very beginning of the process to divide time between analysis and additional data collection. Scheduling time for team interaction is a way to insure team interaction.

> RQI is divided between time for collecting information and time for analysis of information.

RQI is divided between scheduled blocks of time used for collecting information and blocks of time during which the team engages in data analysis and considers the next round of data collection. Even as the team reviews the data collected up to that point and begins to consider possible conclusions, the team makes conscious decisions about additional methodology and lines of inquiry. Specific decisions will be made on questions to revise, add, or delete; methods and techniques to change; and locations and individuals who need to be visited.

Intensive team interaction during this process allows the team to benefit from the perspectives of the different team members. The chances that important issues will be missed decline as input from different team members increases. The discussions on possible conclusions early in the process ensure that even during a short RQI there is time to try out new lines of inquiry in the field and to test possible conclusions against new data.

While the RQI team is searching for trends, patterns, and opportunities for generalization, the iterative nature of the process allows for the discovery of the unexpected. RQI can be thought of as an open system in which what is learned from feedback is used to progressively change the system. The research effort is structured to encourage the RQI team to rapidly change questions, topics for interviews, sources of data, and direction as new information is gathered. One of the most serious problems with the implementation of RQI is the failure to allow sufficient time for multiple iterations.

Techniques Associated with Iterative Analysis and Additional Data Collection

Structuring the Research Time

Opinions differ considerably on how to structure the time of an RQI, but there is almost universal agreement on the importance of explicitly scheduling time for collecting data and specific time for team meetings to make sense out of the collected data. Regularly scheduled team meetings help build team cohesion and provide opportunities for interaction between the insiders and the outsiders on the team. As noted earlier, the scheduling of time for analysis should start at the beginning of the RQI. Scheduling time for team analysis early in the process is necessary to ensure that there will be adequate time for returning to the field to collect additional information, analyze this new information, and then, based on this analysis, return to the field for even more information.

Beginning on day one, schedule time for team interaction.

Schedules should include blocks of time for analysis before a new cycle of data collection is scheduled to begin. If data collection is a daily process, there is a need for daily blocks of time for analysis. In addition, a longer block of time may be necessary at least once a week to review the overall status of the RQI. These weekly meetings may focus more on the identification of possible conclusions than will the daily meetings, in which the focus may be more on immediate decisions on what additional data to collect and the best strategies for collecting. Sometime in the middle of the process the schedule should include time to prepare for and make presentations to the local people (see below, "Checking Back with Informants"). Large blocks of time will be needed near the end of the RQI to prepare the report and to review the process. The importance of the intensive team interaction before each new cycle of data collection cannot be overemphasized.

While each individual RQI is an iterative process, a RQI can also be part of a larger iterative process in which its results are considered exploratory and subject to revision based on a subsequent RQI, other research, or the monitoring and evaluation of an intervention resulting from the RQI. The RAP Sheet includes scheduling a review of the results and possibly updating the report. (See ch. 6, "The RAP Sheet.")

Debriefing the Team

In addition to team meetings following each data collection activity, it may be useful to have a nonteam member, such as a trainer or someone with more advanced expertise in interviewing, debrief the team. Experience reported by an experienced user of RAP has shown that team members want to discuss their experience immediately following a group interview. When given the opportunity to talk about their experience team members can be expected to share their enthusiasm, discouragement, trepidation, and other emotional responses to doing the interview.

Debriefing sessions can be used to keep the momentum and morale up as the team works through difficult circumstances. Debriefing sessions can be with individual team members or the entire RQI team. Debriefing sessions may be especially critical when there are delays in getting the entire team together following interviews.

Checking Back with Informants

If possible, before the conclusions are final, the RQI team should share them with the people who have provided the information and check for agreement. This process is called "member checking." This can be done either formally or informally, but the purpose should be made clear to the local people. The local people can provide corrections to facts and their own interpretation of the situation. Preparation for a presentation of results provides the RQI team with the opportunity for intensive interaction concerning emerging conclusions and gaps in the data. Even before the presentation is made, the research agenda can be expected to change. Team presentation of the tentative results allows for a division of labor, with one or two members focusing on the presentation and the other members focusing on the response, both verbal and nonverbal, from the local people. Ideally, team members who are not presenting should be sitting with the local people. Informal comments from the local people at this stage can be especially useful. A part of the presentation should be to ask advice on how the team might focus its remaining time, with attention to who else should be interviewed. In addition to its role as a research technique associated with the iterative process, checking back with the local participants is a critical part of the data analysis process. The role in data analysis of checking back with the local participants is discussed in the section on techniques for data analysis, below.

> Share tentative results with local participants before the conclusions are final.

RQI Report Preparation by the Team

The joint preparation by the entire RQI team of the RQI report continues the intensive team interaction. The preparation of the report should start while there is still time for additional data collection. Presentation by team members to each other accelerates the analysis process and allows input on possible conclusions from different disciplinary perspectives. Gaps in information that may not be apparent to one team member are unlikely to escape detection by the entire team. The involvement of the entire team, including the local members, in the report preparation provides the report with a level of cross-checking that is impossible with reports prepared by a single individual.

> The entire RQI team should write the RQI report.

The RQI report should be written using a vocabulary and style that is most readable to the intended audience with the lowest level of formal education. The team is cautioned to avoid "Greek-fed, polysyllabic bullshit" (Becker 2007, 10). Becker (2007) provided excellent advice on overcoming writer's block, writing and revising (again and again), and adopting a persona compatible with lucid prose.

Sometimes an RQI report can have greater impact if it is not the traditional, narrative report. Sobrevila, an ecologist with the Nature Conservancy, directed a Rapid Ecological Assessment team whose end product was color-coded maps of an area's vegetative cover. A team of local and Nature Conservancy biologists spent two to three weeks exploring an area and identifying flora and fauna. All of this information was compiled in maps showing different classes of vegetation and estimating the wildlife potential. Sobrevila stated, "We believe maps are a strong conservation tool" (as cited in Abate 1992, 486).

Data Analysis

There are numerous ways of analyzing qualitative data. There is no one best way and researchers often are informal in their approach to analysis. Miles et al.'s (2014) comments about the research process appear especially relevant to analysis: "To us it seems clear that research is actually more a craft and sometimes, an art than a slavish adherence to methodological rules. No study conforms exactly to a standard methodology; each one calls for the researcher to bend the methodology to the peculiarities of the setting" (7).

There is no one best way for analyzing qualitative data.

An approach to analysis that has worked for me is based on the Miles et al. (2014, 12) model. This model of analysis involves three steps: (1) data condensation including **coding** the data and adding marginal remarks, (2) displaying the results of the data condensation, and (3) drawing and verifying conclusions. Figure 4.1 illustrates the relationship of these aspects of analysis and the relationships of the analysis to data collection.

Other qualitative researchers (see Richards 2009; Wolcott 1994) use different categories and a different vocabulary to describe a similar process. For qualitative research, data analysis is an ongoing process that can begin before any data is collected (for example when the illustrative topics and question guidelines are being prepared) and continues through the preparation of the final report.

Start the data analysis process before you end data collection.

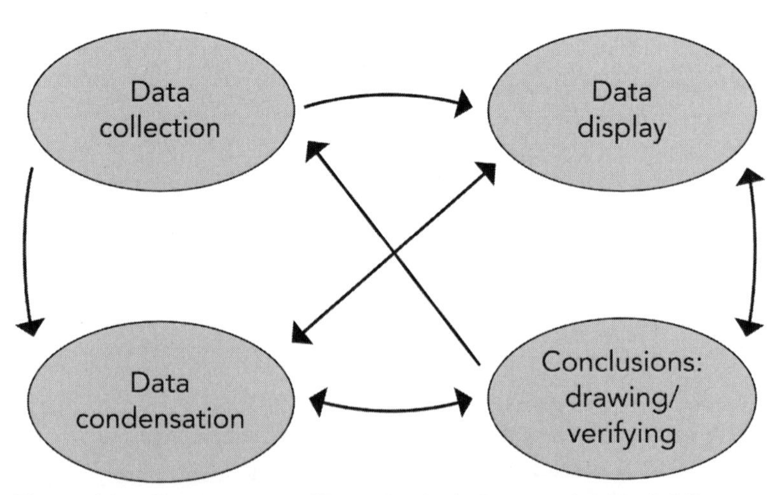

Figure 4.1. Components of Data Analysis: Interactive Model (*Source:* Matthew B. Miles, A. Michael Huberman, and Johnny Saldaña, *Qualitative Data Analysis: A Methods Sourcebook*, 3rd ed., p. 14, © 2013 by Sage Publication, Reprinted by Permission of Sage Publications, Inc.)

Data Condensation, Coding, and Margin Remarks

As described by Miles et al., data condensation

> refers to the process of selecting, focusing, simplifying, abstracting, and/or transforming the data that appear in the full corpus (body) of written-up field notes, interview transcripts, documents, and other empirical materials. (Miles et al. 2014, 12)

The logs are the source of the data for beginning the process of developing an understanding of a situation (see ch. 3, "Managing Data—Field Notes and Logs"). The first step in the analysis process is to read the logs. Logs are read, reread, and reread again. Several authors recommend that logs be read a minimum of three times before anything else is done. Logs need to be reread again before additional data is collected. The next step in the analysis process is dividing the log into thought units and applying codes to these units. A unit of thought may be a sentence, paragraph, several paragraphs, or even an individual word. Coding is the key to the process of selecting, focusing, simplifying, abstracting, and transforming the data that appears in the logs. Coding, like **data display** and conclusion drawing, occurs continuously throughout the research. It may even start before any data have been collected, when decisions are made on research questions, specific research techniques, whom to interview, and conceptual frameworks (Miles et al. 2014, 10). Coding is often the most time-consuming aspect of the data analysis process.

Coding can be thought of as cutting the logs into strips and placing the strips into piles. The codes are the labels you give the individual piles. Some researchers physically cut the logs into strips or use qualitative computer analysis software such as NVivo to divide the log into units to which codes can be applied (see Bazeley and Jackson 2013). In many cases and for many researchers, writing codes in the margins of the logs is sufficient for analysis. I have found writing codes in the margins of the logs adequate for RQIs.

Developing a coding system is based on trial and error and more trial and error. It is critical that the coding system remain flexible. The persons assigning codes are looking for threads that tie together bits of data. They are looking for recurring words or phrases. These words often become the labels for the codes. Many experienced researchers suggest having only five or six major codes and, where necessary, dividing these codes into subcodes. Coding can be thought of as winnowing, since not all information may be relevant or have to be coded. You can always return to the log and code material that was not coded and change the coding done earlier. It should be noted that a single unit of thought will often have multiple codes.

Start with only five or six codes. Subdivide these when necessary.

Entire books and chapters have been written on coding and data analysis, with each suggesting a somewhat different approach (see Saldaña 2013; Miles et al. 2014). Marshall and Rossman's (2011) chapter on "Managing, Analyzing, and Interpreting Data" provides a succinct introduction to these topics that will be especially useful for the less experienced researcher. I have found that the best way of learning to code logs is practicing the process in small groups of two or three people.

Time spent coding logs can help prevent jumping to premature conclusions.

Adding **margin remarks** is closely related to the coding process. Margin remarks are usually written into the margin of the log after it has been typed. Almost anything can be included in a margin remark, but often they are related to the coding activity. Margin remarks made during the coding process include ideas and reactions to the meaning of statements that may not be consistent with codes. Sometimes, margin remarks suggest new interpretations and connections with other parts of the data. Margin remarks can identify themes involving several different codes. Sometimes, trying to differentiate between codes and themes can be a waste of time. Margin remarks may identify issues for the next wave of data collection. Some authors suggest that codes be placed in the left margin of the log and that margin notes be placed in the right margin. Regardless of which side of the log is reserved for codes, everyone who will use the log needs to agree. Margin remarks often include comments similar to some of the comments in MEMOS, with the major difference being that MEMOS are usually typed and margin remarks are usually handwritten and added after the log has been prepared. Figure 4.2 is an example of a log based on several different interviews conducted at different times and includes a MEMO, codes, and marginal notes.

Data Display

The second aspect of analysis is data display. Being the second aspect of the analysis process should not be confused with being the second step.

Codes	Log	Marginal Remarks
SS purpose enjoy S regulations	MG. It is my endeavor here to always serve students and specifically those who need assistance. I can tell you it's not always easy to focus on serving students when you're in the hot seat that we are usually in. We do remind ourselves over and repeatedly -- that is what we are dong here and that is what we are about. It's rewarding and the students are enjoyable. I have worked a very long time; we are forever changing. We have new things thrown at us every year.	
enjoy S tech. S not prep.	KK. I like students. That is why I am here. I've always been in education with regards to hanging around students, and I like that. I don't like touch-tone. I think it is impersonal, cold and half of the time the students don't read the instruction. The students have to come back and see me anyway because they couldn't get registered.	
leadership ↓ resources	AR. (1) It's been a battle in the last seven years because of a particular administrator. . .who didn't really value what we were doing as we would have liked and wanted to do away with position and do a lot more curriculum advisors. The results was that over those year we lot a lot of position due to attrition. This was due to retirement or people taking jobs in other areas. Those positions were not refilled.	downsize be attrition / problem of fairness
tech.	AR. (2) Some things came up, after I talked with you and talked with the counselors, that I wanted to bring to your attention. One of the big problems that we're having right now is technology. Our technology is trying to match up registration with the Web and with touch-tone and is being hampered because of prerequisite codes.	tech. problem due to ↓ resources
	- - -	
↓ resoruces	There is a push in the whole wide world to go technology. And yet, there is the reality that people are really dragging their feet. . . .We want to help students.	
	- - -	
overwork	Right now, it's just creating a massive amount of work for us that we shouldn't have to be doing.	unfair distribution of work
	- - -	
	[MEMO December 3, JB. This was AR's second interview. He was insistent that he needed more time with the team. He clearly had an agenda of items he wanted to cover.]	

S= Students
SS= Student Services

Figure 4.2. RQI Log Illustrating MEMOS, Codes, and Marginal Remarks

There is no particular order in which the data display and conclusion-drawing aspects of data analysis should occur, and even coding is likely to continue throughout the analysis process. Miles et al. (2014, 10) defined a display as "an organized, compressed assembly of information that allows conclusion drawing and action." They argued that the extended amounts of text found in logs can be very difficult to use for drawing conclusions, since they are dispersed, sequential, poorly structured, and bulky. They argued for better displays of data, including different types of matrices, graphs, charts, and networks. Data displays that compare different cases can be especially useful. The data display should assemble organized information into an accessible, compact form. This allows the analyst to "see what is happening" and either draw conclusions or move on to the next step in the analysis (13). As with data condensation, the creation and use of displays is part of analysis. Figure 4.3 illustrates a data display based on one type of matrix. There are numerous other types of matrices, as well as

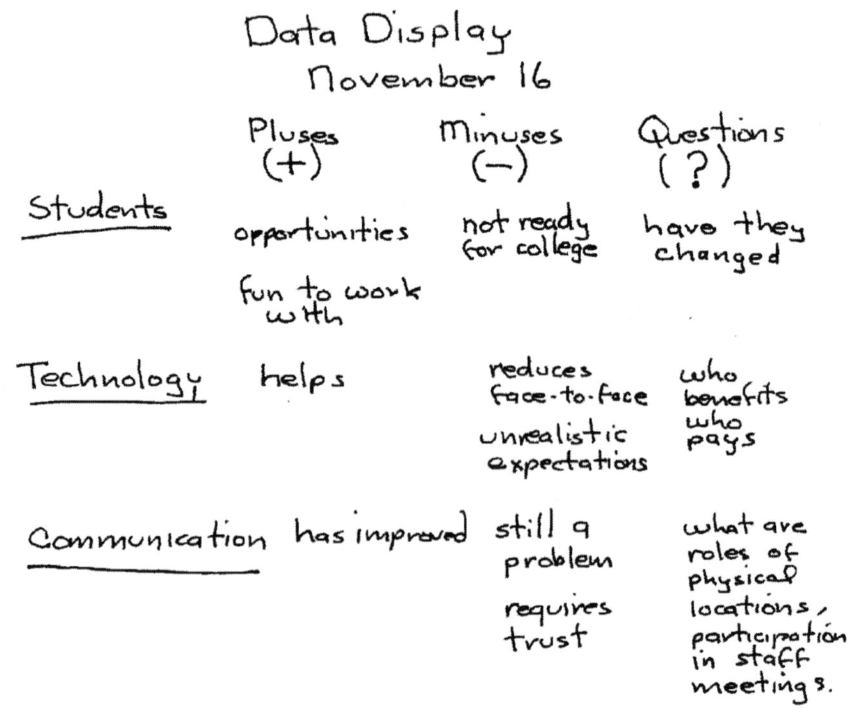

Figure 4.3. Example of a Data Display

different types of graphs, charts, and networks. A rich picture (see figure 3.1) can also be used as a data display.

Conclusion Drawing and Verification

The third element (but not necessarily the third step) of analysis is conclusion drawing and verification. "From the start of data collection, the qualitative analyst interprets what things mean by noting regularities, patterns, causal flows, and propositions" (Miles et al. 2014, 13). Miles et al. suggested the competent researcher "holds these conclusions lightly, maintaining openness and skepticism."

Miles et al. (2014) identified and explained numerous tactics for generating meaning. The first six of these are more descriptive than the others, provide a good beginning point for most research, and will be briefly introduced here. The first three focus on helping to identify "what goes with what" and are concerned with (1) patterns and themes, (2) seeing plausibility, and (3) clustering. The next three tactics for generating meaning, (4) metaphor making, (5) counting, and (6) making contrasts and comparisons, are ways of achieving integration of diverse data and seeing "what's there" (277).

1. *Noting patterns, themes.* Patterns pull together separate pieces of data. Among other things, patterns can be based on similarities and differences or connections in time and space. Miles et al. suggested that patterns occur so quickly and easily that there is no need for how-to advice: They recommended seeking additional evidence for patterns and remaining open to disconfirming evidence (278).

2. *Seeing plausibility.* Sometimes conclusions are plausible, make good sense, or fit and are based on intuition more than anything else: "It just feels right" (278). Miles et al. recommended, "Trust your 'plausibility' intuitions, but don't fall in love with them" (278).

3. *Clustering* consists of "clumping things into classes, categories, bins" (278) and involves putting together things that are like each other. Miles et al. noted clustering provides a sense of

comfort and security. Examples of clusters include roles, rules, relationships, routines, and rituals (279). Clusters may overlap and need not be mutually exclusive. Miles et al. noted clustering is a process of moving to higher levels of abstraction (279).

4. *Metaphor making* is suggested as a way of integrating diverse pieces of data. Metaphors involve comparing two things based on their similarities while ignoring their differences (281). Miles et al. noted that people constantly use metaphors as a way of making sense of their experience (281). Miles and Huberman (1994, 250) suggested that there is a need to be aware of the metaphors used by both the researcher and the people studied and that the researcher should "dig into them for implicit meanings, for more explicit exploration and testing" (282). They suggested asking questions like "What does it feel like?" and "If I only had two words to describe an important feature at this site, what would they be?" as a way of generating metaphors. Follow-up to the reference to "my private prison" by one of the participants in the community college RAP/RQI led to a discussion of physical office space, the participant's perception that she was being observed, and her relationship with her supervisor. Miles et al. cautioned about looking for overarching metaphors too early in a study, since doing so can distract the researcher from fieldwork and lead to hasty judgments. They also recommend knowing "when to stop pressing the metaphor for its juice" (282).

5. *Counting* is suggested as a familiar way of seeing what is there. Miles et al. identified three reasons for using numbers as part of the process for generating meanings. Numbers help the researcher to see rapidly a large amount of data, to verify a hunch or hypothesis, and to stay honest. Miles et al. noted that numbers can be more economical than words (284).

6. *Making contrasts/comparisons* is identified as a time-honored way to test a conclusion by drawing a contrast or making a comparison between two sets of things. Comparisons are

based on asking the question, "How does X differ from Y?" The focus is on the contrast between sets of things, such as persons, roles, activities, and cases that are known to differ in some important respect. Miles et al. recommend asking, "How big must a difference be before it makes a difference?" (284).

The process that has been outlined for analyzing the results of a RQI has included (1) data coding and the related activity of preparing margin remarks; (2) data display, with emphasis on graphical ways of organizing information; and (3) conclusion drawing, with emphasis on six specific tactics for generating meaning. An important aspect of conclusion drawing is verification. According to Miles et al.,

> Verification may be as brief as a fleeting second thought crossing the analyst's mind during writing, with a short excursion back to the field notes, or it may be thorough and elaborate, with lengthy argumentation and review among colleagues to develop "intersubjective consensus," or with extensive efforts to replicate a finding in another data set. (2014, 13)

Without verification, they argue, "we are left with interesting stories about what happened, of unknown truth and utility" (14).

How Much Data Is Needed?

Every qualitative researcher faces the question of how much data is needed, with specific attention to how much time in the field, what number of interviews, and what additional information is sufficient. These are especially important questions for RQI. A qualitative researcher can conclude that her data is sufficient when the collection of additional data does not result in new categories or themes but only new examples (Ely 1991, 158). The criterion of redundancy states that when themes repeat themselves, it is time to stop. Ely suggested, "It is far easier but perhaps more unsettling to know when enough is not enough" (159).

When themes begin to repeat, it is time to stop.

Checking Back with Informants

Checking back with the informants for the RQI is an important aspect of the data analysis process as well as a part of the iterative process of data analysis and additional data collection. A study can be considered trustworthy when the people who provide the information are willing to agree with the analysis of the researchers. Checking back with the local participants lets the RQI team know if participants agree and allows the team to collect additional information and revisit the analysis process if there is disagreement. Checking back with the local Ministry of Agriculture officials during the Polish state farms RAP/RQI was critical for the identification of specific issues that could be addressed at the local level. The officials confirmed that some directors of state farms were using the rhetoric of a market economy but did not really understand the meaning. This became a specific area where the local Ministry of Agriculture could provide support for these directors. The local officials also confirmed the observation that some directors did not recognize that other directors were relatively more successful and that lessons could be learned from them. This became a second area of intervention for the local Ministry of Agriculture. Checking back with the participants in the community college RAP/RQI confirmed the identification of specific areas that needed change and allowed the participants to begin to address some of these issues without waiting for the report. Soliciting the participant's views on the credibility of the findings and interpretations is referred to as "member checking" (Creswell 2013, 252). Lincoln and Guba (1985, 314) claimed member checking was "the most critical technique for establishing credibility."

The next chapter explores the issue of teamwork as an element of RQI, with careful attention to the experience of the Polish state farm RQI. While teamwork is not new to qualitative research, it is not the norm. A bit of background on the use of teamwork should therefore be helpful in understanding the issues faced by the RQI team. Some of the examples discussed in the next chapter will not be completely consistent with the methodology discussed in the preceding chapters. This should not come as a surprise or be a cause for concern. Flexibility and the adaptation to local conditions and resources available are defining characteristics of RQI.

Additional Readings

Miles, Huberman, and Saldaña's (2014) *Qualitative Data Analysis* provides the best model and detailed instructions and examples I know of for data analysis. Ely's (1991) *Doing Qualitative Research* and Marshall and Rossman's (2011) *Designing Qualitative Research* have chapters that provide more general approaches to analysis than Miles et al. Becker's (2007) *Writing for Social Scientists* provides excellent advice on writing, with specific attention to editing to achieve readable prose.

Becker, Howard. S. 2007. *Writing for social scientists: How to Start and finish your thesis, book, or article.* 2nd ed. Chicago: University of Chicago Press.

Ely, Margot, with Margaret Anzul, Teri Friedman, Diana Garner, and Ann McCormack Steinmetz. 1991. *Doing qualitative research: Circles within circles.* Bristol, PA: Falmer Press.

Marshall, Catherine, and Gretchen B. Rossman. 2011. *Designing qualitative research.* 5th ed. Thousand Oaks, CA: Sage.

Miles, Matthew, B., A. Michael Huberman, and Johnny Saldaña. 2014. *Qualitative data analysis: A methods sourcebook.* 3rd ed. Thousand Oaks, CA: Sage.

CHAPTER FIVE

TEAMWORK FOR DATA COLLECTION AND DATA ANALYSIS

Main Points

1. RQI is premised on teamwork and the success of RQI depends upon the quality of the teamwork.
2. The assumption is that the RQI team will be together most of the time and will work together in data collection and analysis, including the preparation of the report.
3. For teamwork to substitute for long-term fieldwork, team members must bring to the team different perspectives and expertise.
4. Several sets of eyes and ears and continuous team interaction are essential for getting the most from the very short time in the field.
5. In both cases discussed in the chapter, by the end of the fieldwork, the teams were spending as much or more time talking with each other and trying to make sense out of what was being observed as they were on observing new things.
6. The ability of RQI to begin the process of achieving an emic understanding of a situation in a short time requires the use of a team of researchers, with, whenever possible, at least one insider on the team.
7. Even when the rest of the RQI team has gone on to other things, the insiders continue to be called upon to clarify results, resolve pending issues, and help organize local responses to the results.
8. Teamwork to some extent depends upon team leadership.

9. The five most critical responsibilities of the team leader are (1) providing orientation, (2) keeping activity focused and responsive to changing conditions, (3) maintaining morale, (4) team building, and (5) ensuring that administrative support is provided.

Teamwork and the Lessons from RQIs

RQI is premised on teamwork, and the success of RQI depends upon the quality of the teamwork. As explained in the preceding chapters, team interaction is critical for both the data collection and data analysis processes. Team interaction depends upon the mix of people on the team and each member understanding her or his responsibility. Whenever possible, the RQI team should include one or more insiders as well as the outsiders. The role of insiders is extremely important and will be given special attention. Finally, the role of leadership in facilitating teamwork deserves close attention. I will draw on the RAP/RQIs of the Polish state farms and a community college to illustrate teamwork. Even though the original documentation on these studies used the label "Rapid Appraisal," the process meets the requirements of RQI and that term as well as the term *RAP/RQI* are used here. The examples are meant to suggest some of the many ways that teamwork can be implemented and are not intended to provide a model for implementing RQI. Because of the extensive use of examples from these RAP/RQIs, a brief introduction for each will be useful.

The Example of the Polish State Farms RAP/RQI

In 1989 the government of Poland, at the end of more than forty years of Communist rule, announced that state industries would be privatized. Inflation accelerated rapidly in 1989 and the economy faltered. On January 1, 1990, the government instituted what was called the "big bang" by decontrolling prices, slashing subsidies, and drastically reducing import barriers. The transition to a market economy proved especially difficult in the agricultural sector. The Communists had consolidated most of the small private farms into collectives, but "state farms," owned and directly managed by the state, controlled 25 percent of the agricultural production. State farms were large enterprises often consisting of several different sites and multiple enterprises including crops, livestock, and processing. They were generally responsible for providing housing and social services like schools and medical care to their workers. Despite the 1989 announcement, none of the state farms had been privatized by mid-1991 and there was growing concern in Poland that many of these farms would fail and their resources would be lost. The government of Poland approached the U.S. Department of Agriculture (USDA) for assistance on this issue. The four American team members arrived in Poland in early September 1991, less than ten weeks after the USDA requested that I organize and lead a RAP/RQI. The American members of the team joined the Polish team member in Warsaw. The entire team spent three days in Warsaw, sixteen days in the province of Koszalin, and four days back in Warsaw working on the report and briefing Polish Ministry of Agriculture officials. The English version of the final report was finished seventeen days later and the Polish version twenty-four days after the English version. The team traveled more than twenty-five hundred kilometers in Koszalin during the course of the RAP/RQI. Team members observed livestock production facilities, fields under regular cultivation, experimental fields, processing plants, apartments, two private farms, and a cooperative store run by a group of private farmers. The team conducted extensive semistructured interviews, averaging more than four hours each with the directors of ten carefully selected state farms. In addition to the ten directors, the team interviewed a livestock unit manager, the chairman of a workers council, the director of a state farm not visited, and a combine driver. Team members participated in meetings with all of the directors of state farms in the region, the mayors of four municipalities, bank officials, researchers from the potato research center in Koszalin, and the director of an agricultural secondary school. More than 110 hours of interviews produced 110 pages of typed logs.

The purpose of this RAP/RQI was to assess management responses to the economic environment with the objective of identifying what, if any, interventions might stabilize the farms until they could be privatized. The team found that the responses of state farm directors to the changes in the economy ranged from no response to changes in organizational structure, crop and livestock production, financial management, levels of employment, and marketing. These responses were analyzed based on the directors' explanations of what had happened. The responses suggested that many state farm directors did not fully understand how a market operates. This lack of full understanding was most serious in areas of (1) risk, (2) information, (3) production versus profits, (4) price determination, and (5) financial (cash-flow) management.

The RQI team identified a role for the Polish Ministry of Agriculture to play in helping state farm managers better deal with the changing economy of Poland. The primary thrust of this assistance was helping directors and others address the issues of risk, information, profits, price determination, and financial management. Specific programs were proposed for (1) helping them locate and interpret market information necessary to predict a range of likely prices, (2) helping them adjust production decisions in response to this range of prices, and (3) helping them improve financial management, beginning with cash-flow analysis, in order to implement decisions based on price estimates. The executive summary for the report is in appendix B.

The Example of the Community College RAP/RQI

As noted in chapter 1, the RAP/RQI at a community college in the Pacific Northwest was initiated by the newly hired dean of the Student Services Division. She understood enough of the depth of mistrust among the employees in her division and the complexity of related issues to know that something other than a traditional approach to understanding these issues was needed. She was aware that not enough was known about these issues to formulate questions that could be used for questionnaire-based research and that there was insufficient time to do traditional qualitative research. She asked me to organize an RQI focusing on the organizational structure of her division.

Most of the employees in the Student Services Division first learned that there would be a RAP/RQI from an e-mail message. While the message outlined the essential elements of the methodology, it did not address the issue of the objectives for the RAP/RQI. It also failed to provide information on who would be participating in the study or the proposed timeline. On the first day, the RAP/RQI team met with the leadership group from the division. At that point, questions were raised about the objectives of the study, how the research team had been chosen, how much control the college had over the RAP/RQI team composition, how much the study was costing, who would "own" the results, and how the data would be used. The team provided straightforward answers to these questions. There was a series of questions about confidentiality and it was stated that some individuals would find it very difficult to participate since they had been harmed in the past by being "open." A joke about the room being bugged elicited a serious discussion about an instance when a past administrator had in fact listened in on a meeting of the administrator's subordinates. The level of suspicion is illustrated by the concern of two participants who did not want anyone concerned with the college to be involved in transcribing their interviews. There was real concern about how choices would be made on who would be interviewed and whether all units and positions would be included. There was some concern about the qualitative process of identifying themes from the information collected. The RAP/RQI team asked the leadership group to inform the individuals in their units about the RAP/RQI and to suggest individuals who might begin the process.

The RAP/RQI did not begin with a preset list of people to be interviewed. Most participants were self-identified or were identified by their colleagues as individuals who should be included. Once a few individuals had been interviewed, there was a flood of individuals who wanted to be interviewed.

The team briefed an expanded version of the leadership group about three-fourths of the way through the process and received feedback on the tentative themes that had been identified. By this point a growing number of individuals still wanted to be interviewed. The RAP/RQI team made a decision to name an arbitrary end point for interviews and to encourage individuals who still wanted to contribute to write their thoughts and submit them to the team. Individuals interviewed included fourteen people from the Student Services Division, one senior administrator from outside of this division, and three students. Almost all interviews were recorded and transcribed. By the time the last interviews were conducted, it was clear that there were major differences in opinions. As noted in chapter 1, one of the significant results of the RAP/RQI was the identification of the differences in how "students" and "services" were defined by the participants. Six constraints were identified that participants felt prevented them from doing as good a job as they would like to. Constraints involved: (1) communication; (2) physical space; (3) technology; (4) utilization of people's time, talents, and creativity; (5) increases in the number and complexity of regulations; and (6) inadequate resources. After the report was completed, the RAP/RQI team was invited back to make a general presentation to almost everyone in the division. Even before the RAP/RQI was finished, the leadership of the community college started to address some of the issues that were identified. After the final report was submitted, teams were organized and funding was made available to tackle specific problem areas. Additional information about the findings of the community college RAP/RQI can be found in appendix A.

Qualitative Research and Teamwork

Most qualitative research is conducted by lone researchers, even though the tradition of team research extends back to the beginning of ethnographic research. If joint authorship of articles is used as a rough indicator of team research, less than 30 percent of research is performed by teams (Erickson and Stull 1998, 5). Most qualitative research and especially ethnographic research has generally been described as an individual undertaking. Van Maanen, Manning, and Miller (1998, vi) used the phrase "one man, one tribe" to describe the tradition, even as they noted the phrase is politically incorrect and historically inaccurate. Guest, Namey, and Mitchell (2013, 12) described the traditional approach as a "lonely enterprise" where the researcher trekked "to some exotic and remote place and lived with the local people for a year or more." In such a situation one individual was responsible for the entire research process.

Erickson and Stull (1998, 60) and Guest and MacQueen (2008, 4) have observed that the trend is for increased use of teams in the conduct of qualitative research. Erickson and Stull suggested qualitative researchers can expect to be on teams with other qualitative researchers and to represent the field of qualitative research on multidisciplinary teams (1998, 60).

As of 2014 there was an increasing body of literature on effective, field-tested, team-based strategies for qualitative research (see for example, Bartunek and Louis 1996; Bradley 1982; Cohen and Bailey 1997; Erikson and Stull 1998; Fernald and Duclos 2005; Marshall and Rossman 2011; Richards 1999). Siltanen, Willis, and Scobie (2008, 58) made a strong case that teamwork "enhances the interpretive creativity" and expands the insights and understanding within a team.

When teams are used to conduct traditional qualitative research, there is often the assumption that team members will be assigned to different sites and be involved in "different arenas of observation" (Erickson and Stull 1998, 18). They referred to a team research effort where team members were rarely together in one place at one time during the two-and-a-half-year study. In contrast, the assumption is that the RQI team will be together most of the time and will work together in data collection and data analysis, including the preparation of the report.

> The RQI team is together most of the time when working on data collection and data analysis.

Erickson and Stull argued, "Teams are not necessarily more efficient than lone rangers in getting their work done, but they can be much more productive" (1998, 36). The increased productivity results from individual team members seeing or not seeing various things and having different interpretations for what they do see (10). Erickson and Stull suggested that it may not be necessary for team members to see the same things at the same time. However, they imply there are advantages for having the team together and doing the observation at the same time (18, 38).

There are some disadvantages to a team approach to research. Given that "the cult of individualism is the state religion of academia" (Erickson and Stull 1998, 54), team research may be more difficult to use for promotion and tenure than research done alone. This will not be an issue for practitioners of RQI outside of academia. Team interaction may make the experience frustrating. Potential sources of conflict include differences in personal style, theoretical and methodological disagreements, divided loyalties, competing professional demands, and unclear lines of authority and responsibility (12). The time team members spend interacting with each other may mean less time interacting with the cultural system they are investigating (55). The researcher involved in traditional qualitative research may have a choice of whether to work alone or as part of a team. This is not a choice for the RQI-based research, since only a team can do it.

CHAPTER FIVE

Teamwork and RQI Data Collection

Putting Together the Multidisciplinary Team

For teamwork to be productive, team members have to be able to bring to the team different perspectives and expertise. In organizing the Polish state farms team and the community college team, I attempted to maximize variability on the teams within the constraints imposed by the availability of individuals to serve on the teams. Based on the expressed needs of the USDA, I concluded that the team in Poland would have to have an economist and an agricultural production specialist. As I started the search for individuals who would be available to participate on the team, I became increasingly aware that the team would benefit from an administrative-organizations specialist, especially someone familiar with Polish culture. I suggested to the USDA that the Polish member of the team should have a relevant technical background. They found an agricultural engineer who had experience with state farms.

For the community college RAP/RQI, my objective was to recruit a multidisciplinary team whose members were at least familiar with the basic concepts of RAP/RQI. All three team members were students in the doctoral program in leadership at Gonzaga University at the time of the study. Two of them had already taken my qualitative methods course. The first team member had a master's degree in engineering, a faculty appointment in the School of Business, and extensive experience in the private sector dealing with information technology and systems analysis. The second team member had a master's degree in nursing, a faculty position in the nursing program, and several years of experience with program administration and higher education administration in general. The third team member had a master's degree in counseling psychology, had prior experience with student services in higher education at another institution of higher education, and had been employed for four years by the community college where the RQI was done.

Diversity

Erickson and Stull noted the importance of ethnic, gender, age, and background differences in teams for traditional qualitative research. They stated that this diversity simultaneously intrudes on and enables the team's efforts to "listen and observe carefully" (Erickson and Stull 1998, 35). They claimed that without gender diversity teams may miss or misinterpret much of their fieldwork (40). Since the RQI team spends less time in the field and has significantly less time to establish rapport, the diversity of the team can be critical in building linkages.

For the Polish state farms study, different individuals on the team interacted with each other with varying levels of comfort. In some cases a shared professional vocabulary was more important than language fluency. Age similarities facilitated communication, especially with individuals who were either particularly junior or senior. The shared status of being professional women in a male-dominated sector provided a special link between the female member of our team and the woman from the Ministry of Agriculture who was a part-time member of the team. The latter accompanied the team for many of the visits and provided liaison between the team and the ministry. The differences in status that Polish officials ascribed to different team members appeared to facilitate their communication with these individuals.

The community college RQI team members were well known to each other and enjoyed working together. As noted earlier, two of the team members had been introduced to RAP/RQI in a qualitative research course taught by the team leader. Familiarity with each other and with the approach facilitated teamwork. One of the team members commented that she was pleased "with how well the RAP team worked together. Each member's skill level was readily apparent. The chemistry of the group was amazing." From the first interviews, there was comfort with team interviewing. All members of the team shared the role of team spokesperson and different members made presentations. The fact that all three team members had experience with community colleges facilitated establishing rapport with numerous participants. The community college RQI team was made up of three women and a man, and included an Asian American woman.

The experiences of the teams in both Poland and the community college suggest that in some cases diversity is critical for enabling a team to proceed rapidly. However, the experience also confirms the observation of Erickson and Stull that the ethnicity of the fieldworkers and the diversity within the team is not always sufficient to guarantee easy access (1998, 43).

Outsiders and Insiders in Between

Van Maanen et al. (1996) argued, "Any social group deserving of a label is one in which members in good standing are able to distinguish insiders from outsiders" (v). These authors noted that social research is based on efforts by investigators who have traditionally been outsiders trying to understand what the insiders believe, value, practice, and expect. While traditional research methodology has focused on helping the outsiders better understand the insiders' knowledge, there is a growing realization of the role the insiders should play in the design, implementation, and publication of research.

Van Maanen et al. (1996) also noted that boundaries may be permeable, groups may overlap, and status can be relative, shifting, and ambiguous. This lack of clear definitions also applies to the identification of who is an insider for purposes of inclusion on the RQI team. It is desirable for the insider to be a member of the local group that is the subject of the RQI. However, sometimes the insider's role will be filled by someone from a neighboring organization or village who speaks the local language. In addition to language competency, at a minimum the insider needs to have had experiences similar to the ones most relevant to that being explored by the RQI.

The ability of RQI to even begin the process of achieving an emic understanding of a situation during a short period of time is significantly improved if the team of researchers includes at least one team member who is an insider. Bartunek and Louis (1996) described research conducted collaboratively by insiders and outsiders as involving the insiders in "examining" the setting and coauthoring any public accounts. They noted that this includes having the insider involved in framing questions that guide the study and playing a role in the analysis of the data. Bartunek and Louis claimed that "by capturing, conveying, and otherwise linking the perspectives and products of inquiry of both insiders and outsiders, a more robust picture can be produced of any particular phenomenon and setting under study" (11).

> Whenever possible the RQI team should include an insider.

Bartunek and Louis further argued, "The more diverse the experience histories of the individuals composing a research team, especially in terms of their relationship to the setting, the more diverse should be their perspectives on and potential interpretations of any particular observed event there" (18).

The role of the insider on the RQI team is made more complex by the "action" objective of RQI of facilitating change in the setting under study. Bartunek and Louis (1996) suggested that where research has practical as well as scholarly ends, the insider can be expected to be motivated to improve his or her lot. They go on to say,

> If the outsider "goes native" or the insider "goes stranger," however, gains previously possible are compromised. . . . In identifying with insiders, outsiders may be tempted to gloss over controversial issues in the setting that are pertinent to the study. In identifying with outsiders, insiders may be tempted to accept uncritically outsiders' categorizations for behavior, rather than contributing their own unique perspective. (56–57)

Bartunek and Louis (1996) identified several ethical challenges in conducting team research using both insiders and outsiders. **Informed consent** (discussed in more detail in chapter 8) is made more problematic, since prospective participants cannot have full knowledge of what might come out of the research. The inability to know in advance what might develop is especially a problem for the insider member of the team. Bartunek and Louis argued, "A revised view of informed consent seems warranted, in which consent is negotiated at different points in the research cycle" (58). Confidentiality represents another ethical dilemma that is particularly relevant in presenting results of studies that include insiders as team members. Even when pseudonyms are used, there is a living link between quotes and individuals. Where the decision is made not to identify the site of the study, the institutional affiliation of the insider can sometimes identify the site.

Initial plans for the Polish RQI did not include explicit reference to the involvement of insiders. According to Bartunek and Louis (1996) this is not unusual for teams that eventually include insiders. The Polish agricultural engineer was clearly an insider for the agricultural sector of Poland. While he had experience with state farms, he was not an insider for the state farms in Koszalin province. From the initial meeting with the American team, he was considered a full member of the team. Urszula Golebiowska, the chief specialist for economics and transformation in the

Ministry of Agriculture for Koszalin, was very much an insider to the state farm community in Koszalin. She joined the team for almost all of the visits to the state farms. In addition to providing introductions for the rest of the team, she answered questions from the other team members and ensured that the team did not overlook critical issues.

The demands of his job prevented Jansz Turski, the director of the Koszalin office of the Ministry of Agriculture, from participating in as much of the team efforts as he would have liked. His contributions were especially significant at the beginning and the end of the fieldwork. At the beginning of the work, he suggested areas of concern and identified opportunities for the team to participate in events that were already planned, such as a meeting of all the directors of state farms in the region. His greatest impact, however, was at the end of the process. The rest of the team benefited from being able to explore preliminary ideas about areas where state farm directors had the greatest misunderstanding of a market economy. Turski communicated to the rest of the team the need to identify specific recommendations and to clearly define an explicit role for the Ministry of Agriculture. He was responsible for arranging meetings between the team and the leadership of the local ministry and provincial officials. These meetings were critical for the productivity of the iterative process.

The role of the insiders in this RQI activity contributed to the ability of the team to produce the timely results that had been requested. Because of their self-interest in the results, the insiders helped ensure that the results made sense and that recommendations could be implemented. Finally, their presence helped with the implementation of the recommendations. The ministry used the results to begin refocusing their training. They also used their involvement in the research to facilitate the assignment of two USDA agricultural systems specialists to the Ministry of Agriculture.

The community college RQI was designed to include an insider as a full team member. She was the first team member identified and she played an informal role in the selection of the other members. Because of her longer experience at the college than the new dean, she was in a position to provide advice on the types of individuals who would be most appropriate. Her status as a student in the same doctoral program as the other team members contributed to her acceptance by the team as a full member. Because she was obviously respected by her community college colleagues, she was able to allay their concerns about confidentiality and reinforce the rationale for the study. Participants were explicitly asked whether they had concerns about her presence during the interviews; no one did.

Despite trying to remember which "hat" she wore at work during the RQI and to act accordingly, some of her colleagues refused to differentiate between her as "the faculty member" and her as "the RAP team member." Some of the demands of her colleagues caused role conflict. She wrote, "For example, I was perceived as being 'favored' by the dean simply because I was chosen to work on the study. Given my 'favored' status, a few assumed I knew more about the dean and her motives than anybody else. 'How well do you know her?' I was asked on more than one occasion." When it became obvious that not everyone wishing to be interviewed could be, she was the one who explained the situation and listened to the complaints.

Her role as the insider caught between the local situation and the outside team is best illustrated by her involvement with the two participants who wanted to be interviewed but were extremely concerned about the possibility of retribution. They approached her, asking to be included. She then had to reassure them about what would be done to protect their confidentiality, while at the same time cautioning them that, although the team would make its best effort, confidentiality could not be "guaranteed." She listened to their concerns about having the transcripts of their interviews transcribed by anyone associated with the college, and the team made arrangements to have them transcribed by someone not associated. She provided input to the rest of the team on the materials from their interviews that were so specific to their situation as to make their source obvious. When much of these two participants' concerns were not included in the report, it was the insider on the team whom they approached with a request to have information shared with the dean, but not included in the report. The insider negotiated the arrangement between the dean and the team for an oral presentation focusing on the concerns of these two participants. The insider understood their sense of urgency and made sure that the meeting occurred. She also made sure that the two participants knew it had happened.

> After the rest of the team has moved on, the insider still has to deal with the consequences of the RQI.

Even when the rest of the RQI team had gone on to other things, the insiders continued to be called upon to clarify results, resolve pending issues, and help organize local responses to the results. The community college insider was asked to predict the outcomes of impending changes in the organization structure and to "read the mind" of the dean about critical issues. She was told information that her colleagues implied she should "share" with the dean. She wrote, "When I pointedly sat silent with no offer to run to the dean with what I was being told outside of the RQI interviews, I was met with looks of disappointment, as if I were committing betrayal." She was also aware of avoiding situations where she might be perceived "as taking advantage of information gained through the RAP process." Despite what she identified as the "cultural land mines" resulting from her role as an insider on the RAP/RQI team, she was emphatic that she would repeat the process if given the opportunity.

Team Dynamics

In contrast to traditional qualitative research, in which team members are often assigned to different sites and involved in different observations, RQI is based on most of the team being together most of the time. The diverse, multidisciplinary team is the major source for triangulation of data. The assumption is that given the differences in backgrounds, different team members will be seeing and hearing different things. When visiting different swine production facilities in Poland, the production specialist on the team noted differences in the feed. This in turn allowed the team to begin exploring the differences in the way farm managers viewed and were using information. Questions about large, aboveground pipes led to an understanding of the relationship between state farms and the workers who lived in apartments on the farms, and for whom the pipes provided heat from a central power source. The economic collapse of the farms would literally turn off the heat, with devastating results for workers and retired workers who lived on the farm.

The economist, production specialist, and systems specialist were all bothered by answers from some managers to questions about production that on the surface did not make sense. Answers to follow-up questions and subsequent analysis of these answers provided the beginning point for understanding that some of the managers were confusing production and productivity. For each interview, the team would agree on who would take the lead in asking questions, but everyone was expected to participate in every interview.

A major issue at the community college was problems with technology, including the lack of technical support. One of the RAP/RQI team members' fluency in the vocabulary of computers facilitated these conversations. Discussions of the special needs of international students were facilitated by the international student experience of the team leader as a graduate student in the Philippines. The significance of having four or five signs pointing the way to the new office of one of the administrators likely would have been missed by most researchers. The physical presence of a team of researchers in the location and interviewing some individuals in their offices resulted in one member commenting on the number of apparently new signs for one particular office. This led to a discussion of the reasons behind the move of this administrator, which in turn led to a rich discussion on communication problems.

Both teams discussed in advance the need to prevent respondents from feeling that a gang of researchers was attacking them. The use of an interpreter in Poland for much of the questioning helped ensure that there was time between questions and follow-up comments. The presence of Polish speakers on the team helped the interpreters understand questions involving unfamiliar technical terms and helped to ensure that responses were reported in full. The team decided that two or three of its members could do a better job with some visits and interviews. In one case in Poland, after it became obvious that the team approach to questioning was leaving the respondent uncomfortable, the economist completed the interview by himself.

In Poland the team shared the responsibility for taking notes, since interviews were not taped. From the beginning, everyone knew that field notes would be shared and that everyone present for an interview had a responsibility to pay close attention. In some situations, two team members shared responsibility for note taking. This helped ensure that when team members were involved in leading discussions, they did not have to divide their attention between the interview process and taking notes on the interview. In the community college RAP/RQI, interviews were recorded but note-taking responsibilities were still identified. The notes became critical when it was not possible to get the recordings of three interviews transcribed.

The experiences of the teams in Poland and with the community college convinced the members that some of their concerns about a team approach to data collection were unfounded. Being present for inquiries into areas outside the expertise of individual team members turned out not to be a "waste" of time. In Poland we may not have equally enjoyed trips to the swine production facilities, especially the one in a barn over one hundred years old; however, we all learned and were able to contribute to an understanding of how these facilities fit into the state farm system. Likewise, it was not an enjoyable experience for everyone on the team at the community college to sit through long accounts of problems with communication and the pain experienced by some of the participants. However, it was important to have the perspective of the different team members on this information. The insider on the community college team commented that each person on the team "displayed an admirable level of respect for the people being interviewed. I saw people whose behavior typically is full of hostility and mistrust open up on a real level. . . . What this project provided . . . was hope, for people to have a voice in a large, bureaucratic organization that typically does not honor individual voices, especially those of 'classified staff.'" Members of both teams came to appreciate that several sets of eyes and ears and continuous team interaction were essential for getting the most from the very short time spent in the field. Finally, our initial apprehensions about individual respondents feeling intimidated by a "gang of inquisitors" turned out not to be a problem, except in one case. Our sensitivity to this possibility helped prevent it from becoming an issue and helped us recognize the one situation where it was. Team interaction was absolutely critical for team productivity.

Teamwork and Data Analysis

Iterative Data Collection and Analysis

Teams often met immediately after interviews or other data collection activities or on those days when they did not collect information. These meetings focused on what had been learned, what research strategies seemed to work best, and what should be done differently. In Poland, an effort was made to have the field notes typed each day, and a review of the field notes was often the starting point for the discussion. At the community college, transcripts were used as the beginning point for discussion when they were available. Special attention was given to follow-up questions that needed to be asked and specific responsibility for asking these questions was often assigned.

Teams should meet as often as possible.

The teams paid close attention to contradictions and unexpected responses. During an initial group meeting with all of the directors of state farms in the regions, it was stated several times that all of the state farms were failing. When one of the first farms visited turned out to be doing well, the team began to focus on exceptions and to ask about them. A report on the devastating impact that very high interest rates had on production loans led to an inquiry about the banking system and changes in the sources of finances for state farms. It quickly became apparent that under the old system the state advanced funds when needed and directors did not have to consider cash flow.

In Poland, in addition to the daily scheduled meetings, the team discussed various issues while traveling to and from interviews. After the first week of fieldwork, the team spent an entire Sunday afternoon reviewing the progress that was being made and changes that were needed.

Before the interviews with the directors of the last two of the ten state farms, the team met and made some initial decisions about responsibilities for the preparation of sections of the report. While it was obvious that, with only a few exceptions, state farm directors did not understand how a market economy was intended to operate, it was less obvious whether there were specific aspects of a market economy they did not understand. This was the first of several meetings in which the team explored specific themes. Increased focus on a range of issues during the last few interviews helped the team reach consensus on five areas it identified as problematic for many state farm directors: (1) risk, (2) information, (3) production versus productivity, (4) price determination, and (5) cash-flow management.

The community college RQI team faced different issues requiring different approaches. After the first week of interviews, the team started the analysis process by dividing up the transcripts and letting individual team members begin the coding process. The team then met to discuss and list the codes that individuals had used. As a group, the team considered possible themes and developed several data displays that considered possible relationships of the themes. The first several interviews confirmed the impression of deep distrust evident in the initial meeting with the leadership group from the Student Services Division. The team discussed strategies for exploring the extent to which the distrust was related to personality differences as opposed to structural issues. The interview/topic guidelines were modified to ensure that information was collected on the length of service of individuals, the different roles they had held, and their experience with other institutions of higher education. From the first several interviews it was obvious to the team that there were serious differences in how individuals viewed the mission of the division and that some of these differences were related to differences between the unit responsible for general counseling and units responsible for providing services to students with special needs. The team considered ways of understanding these differences without making the situation worse. One response of the RAP/RQI team was to place more emphasis on getting participants to identify positive aspects of their work environment (see ch. 1, "RQI and Appreciative Inquiry"). The community college RAP/RQI team faced three related problems that had to be discussed during the group meetings. The team believed that arrangements were in place to ensure that transcripts of interviews would be available the next day. After the first few days, the turnaround time for transcripts moved from overnight to several days. At the same time, the number of individuals who wanted to be interviewed increased dramatically, with several individuals requesting the opportunity to meet a second time with the RQI team. Delays in the production of transcripts and scheduling additional interviews pushed the process beyond the original schedule and team members began to run into schedule conflicts involving other things that they had planned prior to beginning the RQI.

In both cases, by the end of the fieldwork, the teams were spending as much or more time talking with each other and trying to make sense out of what was being observed as they were on observing new things. However, these discussions were interspersed with additional data collection. The teams used the additional data collection to resolve questions. The diversity of the teams and their ability to bring different disciplinary perspectives may have been more important for the iterative data analysis process than it was for the data collection process. By the end of the RAP/

RQI, there appeared to be genuine respect for the different technical expertise that team members brought to the discussion.

Report Preparation and Additional Data Collection

The report preparation provided additional opportunities for team interaction and the identification of needs for additional data. Reviews of early drafts of the written sections by other team members led to corrections and new insights. In the Polish RAP/RQI, writing responsibilities were changed in two cases. Members of the community college RQI team distributed drafts to the other team members as e-mail attachments and, for making suggestions for changes in each other's work, used the "Track Changes" function in MS Word.

Even as the Poland team continued to gather data, they met with the local Ministry of Agriculture. At this meeting, the team presented their initial findings. There was general support for the findings regarding areas where directors of state farms misunderstood the requirements of a market economy, but open disagreement with the some of the initial recommendations of the team about how to address these issues. The ministry's concerns were worked into the report and the recommendations were modified.

Before the team left Poland, the English version of the final report was in draft. This version was left with the Polish member of the team so that he could start a translation into Polish. Once the English version was final, it was sent to the Polish member so that he could modify the Polish version to make it consistent with the English version. The Polish-speaking American member of the team reviewed the Polish version before it was released. Special effort was invested in trying to produce a report with a consistent message and ensuring that the English and Polish versions were similar in content. Gow (1991) argued that "a team that speaks with more than one voice is doomed" (12). While "doomed" may be an overstatement, the RAP/RQI team for the Polish study believed in the merit of trying to have the report speak with one voice, and made sure that adequate time was invested to realize this goal.

As noted above, the community college RAP/RQI team was not able to complete the final report as scheduled. Additional discussion of the factors that contributed to the delays is in chapter 10. This delay impacted the results of the RAP/RQI in at least three ways. As soon as the presentation on the initial findings had been made, but before the final report was submitted, the division began to address some of the issues. By the time the final report was submitted, a plan was in place for organizing groups around the constraints identified by the report and limited funding was made available to the groups. The RAP/RQI team recognized that changes in the wording of the report could contribute to this process without violating the integrity of the report. While the substance essentially remained the same, the tone became more positive. Finally, the delay gave the RAP/RQI team time to consider the request for interviews by two participants who had a very serious conflict with their supervisor. They indicated they had been especially concerned about confidentiality and feared repercussion. At the same time, they insisted on being interviewed. They made it clear to the "insider" on the team that they wanted input on the report. The RQI team concluded that if much of their input was included, they would be very easily identified, and that their issues were somewhat different from the issues of the other participants. A solution negotiated by the "insider" was a special presentation by the RAP/RQI team to the dean on their issues concerning their superior while excluding this from the report.

The explicit division of time between data collection and data analysis from the beginning of the fieldwork ensured that a mechanism was in place for sharing observations and interpretations. Clear and complementary responsibilities for team members, and mechanisms for sharing observations and interpretations, were identified by Erickson and Stull (1998) as necessary for groups of researchers to function as a team. The structure of RQI ensures that these requirements are satisfied (see ch. 3, "The Use of Teams").

Team Leadership

RQI depends upon teamwork, and teamwork to some extent depends upon team leadership. In what I view as a very insightful observation, MacQueen and Guest (2008, 3) proposed that successful qualitative research "requires an ability to lead and be led." In the words of Fernald and Duclos (2005, 360) the team leader "sets the tone for how the team can work together." I have witnessed the situation described by Erickson and Stull and have seen the frustration it can produce:

> Administrative inexperience, often combined with reluctance to lead, are common problems for research teams. Academics by and large do not like structured work settings and often rebel against hierarchy. (Erickson and Stull 1998, 30)

Teamwork depends on leadership.

Leadership should not be confused with either the exercise of control or the use of power. Heifetz discussed an approach to leadership that is especially relevant to RQI. He called this approach "adaptive leadership." The first objective of the adaptive leader is to give individuals involved in a complex situation the opportunity to participate in finding solutions to their own problems (Heifetz 1994, 85). Heifetz identified several other principles of adaptive leadership relevant to RQI, including providing "holding environments" (66), identifying adaptive challenges, and keeping distress within a productive range (207). According to Heifetz, the adaptive leader uses authority to construct relationships in which the participants raise, process, and resolve tough questions for which there are no obvious answers (85).

RQI teams often must be more structured than the loose alliance model that has characterized traditional team-implemented qualitative research. Under the loose alliance model, fieldworkers may work on individual projects and may write up findings separately (Erickson and Stull 1998, 14).

Team leadership is especially important for structuring teams and maintaining the structure. The relative status of team members can undermine team structure. Erickson and Stull cautioned that team members of similar professional status and age may balk at formalized organizational structure and defined roles (15).

The extent of prior experience with teamwork needs to be considered in determining how much structure is needed. Teams with less experience might do better with more directive leadership. Fernald and Duclos (2005, 361) urged team leaders to be sensitive to behaviors that indicate the team is not working well together such as silence, outbreaks, and missed deadlines and for the leader to be prepared to intervene. The key is that there must be some clear sense of organization.

I am convinced that the five most critical responsibilities of the team leader are (1) providing orientation, (2) keeping the activity focused and responsive to changing conditions, (3) maintaining morale, (4) building the team, and (5) ensuring that administrative support is provided.

Orientation for RQI requires convincing the team that the success of the effort requires close teamwork to compensate for the lack of time in the field. The difference in the amount of time spent on a RQI relative to a traditional study can be shocking to traditional researchers. One member of the Polish team suggested that he had spent a longer time designing his last study than the RAP/RQI team spent doing the study and completing the report. For team members without formal training in qualitative research, the most important message is that the objective in interviewing people is to get them to tell stories and not just provide answers to questions. Orientation for the language interpreters involved both telling them what was expected and then practicing with team members. Anytime the translation seemed shorter than the original response we were suspicious. The Polish-speaking American team members were given the responsibility of ensuring that rich responses survived the translation process (see ch. 10, "Orientation").

The success of RQI depends upon maintaining balance between focus and flexibility. The team leader has a special responsibility for ensuring that interview guidelines and critical issues are not ignored, unless the team makes an explicit decision to change them. However, the team leader shares responsibility with the rest of the team to ensure that unexpected issues that come up receive the attention they deserve. For the Polish RQI, it was my responsibility to ensure that the success stories from some of the state farms were kept in perspective and that the team did not spend a disproportionate amount of time investigating some fascinating efforts to develop tourist facilities that had little relevance for most state farms in Poland. Responsibility for maintaining balance between focus and flexibility was shared by the entire team for the community college RQI.

The team leader responsibilities for maintaining morale and team building are related. For the Polish RQI, morale depended upon moderating the pace of activities before the team killed the leader. Scheduling time for team members to be away from the team and arranging some work-related tourist events helped. Thompson (1970) stated, "My most constant and difficult role turned out to be that of morale builder. . . .

Someone has to keep up the morale of the group" (60). Differences in experience and status imposed a special burden on the team-building effort. Even if more experienced team members value the contributions of less experienced members, the less experienced members may doubt their contributions. An important aspect of team building was communicating the message that the contribution of all team members was valued.

To ensure that maximum time is available for data collection and iterative data collection and analysis, logistics have to work (see ch. 10, "Logistics—Keeping RQI from Becoming SAP"). Someone has to arrange for vehicles, housing, and meals when the team has to travel to conduct the RQI. The Polish RAP/RQI team brought their own laptop computer, but local arrangements had to be made for a printer and paper. One of the major issues the community college RQI team faced was delay in the preparation of transcripts of the taped interviews. I will return to this logistical problem in chapter 8, when I discuss the responsibilities of the sponsors versus the responsibilities of the RQI team.

The leader is responsible for ensuring that the administrative needs of the team are met but does not necessarily have to do it him- or herself. The Polish team was fortunate in being able to arrange for the services of an individual who did an excellent job in this area.

Teamwork and the Success of RQI

RQI depends on intensive teamwork to compensate for the short amount of time spent on fieldwork. The experiences of the teams in Poland and at the community college suggest that this can be accomplished successfully. There are numerous issues that influence the productivity of teams. Among the more important issues are the organization and structure of the multidisciplinary team, the team leadership, and the roles of insiders on the team. For team-based research it is especially important to leave as little as possible to chance and to document processes (Fernald and Duclos 2005, 363).

The next chapter explores issues concerning the extent to which RQI can be trusted. Some of the charges against RQI can be answered, but there are other concerns that are not so easy to deal with. Keeping RQI flexible while also keeping it rigorous is an issue that can be addressed. No one should trust the results of an RQI unless the local people trust the RQI team. At the end of the next chapter, I will return to the issue of bad RQI and good RQI, and you would be disappointed if I did not raise once more the issue of RQI done too quickly. I will not disappoint you.

Additional Readings

Much of the conceptual material in this chapter is based on Bartunek and Louis's (1996) *Insider/Outsider Research* and Erickson and Stull's (1998) *Doing Team Ethnography: Warnings and Advice.* Heifetz's (1994) *Leadership without Easy Answers* provides an excellent introduction to the leadership style I believe is most relevant to successful RQI. Lassiter's (2005) *The Chicago Guide to Collaborative Ethnography* explores ethnography as a collaborative process where others are engaged in a deliberate and explicit manner at every point in the process. Lassiter's focus is on collaborative relationships between researchers and the object of their fieldwork and not on collaborative relationships among researchers involved in doing the research. Many references to team-based qualitative research concern large, multiple-site, multiyear projects. The most comprehensive reference for large-scale, team-based research I know of is Guest and MacQueen's (2008) *Handbook for Team-Based Qualitative Research.* Even though the focus is on large-scale projects, there is useful information for small team-based efforts such as RQI. Fernald and Duclos's (2005) article "Enhance Your Team-Based Qualitative Research" provides suggestions for improving team-based qualitative research, but with a focus on large-scale projects.

Bartunek, Jean, and Meryl Reis Louis. 1996. *Insider/outsider research.* Thousand Oaks, CA: Sage.

Erickson, Ken C., and Donald D. Stull. 1998. *Doing team ethnography: Warnings and advice.* Thousand Oaks, CA: Sage.

Fernald, Douglas H., and Christine W. Duclos. 2005. Enhance your team-based qualitative research. *Annals of Family Medicine* 3(2): 360–64.

Guest, Greg, and Kathleen M. MacQueen. Eds. 2008. *Handbook for team-based qualitative research.* Lanham, MD: AltaMira.

Heifetz, Ronald A. 1994. *Leadership without easy answers.* Cambridge, MA: Belknap.

Lassiter, Luke E. 2005. *The Chicago guide to collaborative ethnography.* Chicago: The University of Chicago Press.

CHAPTER SIX

CREDIBILITY AND TRUSTING RQI

Main Points

1. At the heart of the question on whether to trust the results of an RQI is deciding whether RQI is an appropriate methodology for a specific situation.
2. For most situations, RQI should produce a sufficiently rich understanding of the insider's perspective for the design of additional research or for initiating activities that need to be started promptly.
3. Spending too little time on the RQI is the most serious threat to its rigor. If done too quickly and without sufficient methodological rigor, RQI can be more dangerous than "research tourism."
4. Gender diversity may be vital in establishing the credibility and trustworthiness of the team with community members.
5. The RQI team should seek out the poorer, the less articulate, the more upset, and those least like the members of the RQI team, and it should involve them in both data collection and analysis.
6. Usually, rapid research methods should not be used for estimating numbers or percentages of a population with specific characteristics.
7. Calling research methods rapid has been used to justify and legitimize sloppy, biased, and rushed work.
8. Flexibility is critical to making RQI relevant to a wide range of systems and is a major strength of the approach. However, this flexibility can be abused.

9. Use the "RAP Sheet" to document what was done and to allow the reader to judge the quality of the work.
10. RQI/Mini-RAP can provide students doing short-term research activities for courses an introduction to credibility issues relating to qualitative research.

Trust and RQI

Before anyone embraces RQI as a methodology, she or he needs to consider the following questions:

- Is RQI too quick for doing qualitative inquiry?
- Does RQI reduce support for more appropriate research methodology?
- What are the legitimate concerns about RQI and can they be addressed?
- What can be done to make RQI both flexible and rigorous?
- What makes good RQI into bad RQI?

At the heart of the question as to whether to trust RQI is deciding whether RQI is an appropriate methodology for a specific situation. If a situation is appropriate for RQI, the next questions concern the rigor with which the methodology is applied.

RQI, the Insider's Perspective, and Time in the Field

Almost all descriptions of qualitative research and especially ethnography refer to a requirement for prolonged periods in the field. Wolcott (2005) described fieldwork as intimate, long-term acquaintance (60). He noted that the previous ideal of two years or longer in the field as the standard had been shortened to twelve months, but that few can afford to spend even this much time in the field (69). Wolcott argued that a minimum of twelve months of fieldwork is often needed if one is to be present through a full cycle of an activity (69). He noted that the early anthropologists studied people through the cycle related to the annual growing seasons. The need to gain access and establish rapport has been noted as justification for prolonged fieldwork, as has the need to observe normal situations that are reflective of the everyday life of individuals, groups, societies, and organizations. Bernard (2011) noted that traditionally fieldwork required a year or longer because "it takes that long to get a feel for the full round of people's lives (281).

The downside of prolonged fieldwork can be the volume of data collected. My own experience was that the almost one year I spent in the field for my dissertation research resulted in something that can best be described as "**social science voyeurism**." Social science voyeurism involves making inquiries on human behavior and practices to satisfy curiosity beyond what is needed or can be used.

Despite the widespread identification of qualitative research and especially ethnography with prolonged fieldwork, there are occasional references to the possibility that qualitative research can be done in less time. Anthropologists including Robert Redfield, Sol Tax, James Spradley, David McCurdy, Penn Handwerker, and Margaret Mead have either described or defended rapid qualitative research methods.

There is an unfortunate tendency to equate time with quality and to dismiss quick results with terms like "quick and dirty." Even researchers like Wolcott (2005) with valid concerns about the quality of "quick" results concede that "time in the field is no guarantee of the quality of the ensuing report" (104). The case for prolonged fieldwork is based on the arguments that (1) it takes time to develop intellectualized competence in another culture (Bernard 1995, 140, 150), (2) it takes time to be accepted and to develop rapport with the locals and this is related to being able to cover more sensitive topics (Bernard 2011, 143), (3) it takes time to be included in gossip, (4) it takes time to get information about social change, and (5) it is the traditional way of doing fieldwork (Wolcott 2005). There is a degree of validity to the first four of these arguments. They may, however, be valid concerns for "quick and dirty" methods that do not meet the requirements for RQI.

The argument on the need for time to develop competence in another culture rests on two assumptions. The first is that fieldwork will be done in an exotic location where the researcher does not speak the native language and has not picked up the "nuances of etiquette" from previous experience (Bernard 2011, 143). This may or may not be true for any qualitative research, including an RQI. Even as Bernard argues for extended time in the field, he noted participant observation studies that have been done in a matter of weeks or months or even "just a few days." Bernard identified the Laundromat as a place where one could do "useful" observations in less than a week based on previous experience in Laundromats (261). I would argue that this same argument would hold for public health experts trying to examine factors associated with risky sexual behavior in their own country, educators looking at the organizational culture in a school, or Thai aquaculture experts interested in possible problems with a new way to manage fish ponds. In each of these cases, the individuals may be facing a complex situation in which, despite their knowledge of the local language and etiquette, they do not know the categories used by the individuals most closely involved in the system. When the locale for the research is an exotic location, then the second assumption that the researcher must develop intellectualized competence comes into play. RQI is based on the full participation of local insiders as team members whenever possible, and the assumption is that they bring knowledge of the language and etiquette. The use of the local team members and the careful and thoughtful use of language interpreters as outlined in chapter 3 can help address this problem. Bernard (2011, 264) noted that when Chambers is called on to do Rapid Assessment of rural village needs, he takes the people fully into his confidence as research partners.

The argument on the need for extended time to develop rapport is based on three assumptions that may not apply to RQI. The first is that deep rapport based on extended interaction is necessary to deal with sensitive topics, like sexual behaviors and political feuds. Usually an RQI is done in response to an identified need or problem. Because of the topic focus of the RQI and the shared concern about the topic at the local level, there is usually a willingness to discuss it. My experience has been that cooperation can be expected if local people recognize that outsiders are genuinely interested in working with them to address their problems. Sometimes just a genuine willingness to listen is all that is needed to get cooperation. The second assumption is that it always takes time to develop rapport. My experience is that personalities and shared interests are more important for establishing rapport than the length of time people spend with each other. Diversity on the RQI team increases the possibilities that there will be individuals who by nature are able to develop rapport quickly and that there will be similarities between some team members and local people. Finally, the third assumption is that, without extended periods to develop rapport, the team using a rapid approach is limited to "going in and getting on with the job of collecting data" and that this means "going into a field situation armed with a list of questions that you want to answer and perhaps a checklist of data that you need to collect" (Bernard 1995, 139). RQI explicitly is based on getting more than the answers to questions prepared in advance. The assumption for RQI is that enough cannot be known in advance to formulate these types of questions. The acrimonious relationship between a few anthropologists and the local people, as evidenced by harassment and assaults, suggests that prolonged time in the field does not always equate to the development of rapport (see also Lee 1994).

One of the arguments for extended fieldwork is that it takes time to be included in gossip. The assumption is that especially valuable information will likely come from informal interviews in which informants gossip freely. Anyone who believes it takes a prolonged period to gain access to gossip never had the opportunity of taking an elevator ride with my late mother. The sharing of gossip almost always encourages the production of gossip.

The assumption that without extended fieldwork there will not be an opportunity for interaction between researchers and local people outside the formal research structure does not hold for RQIs. The inclusion when possible of locals as full partners on the team and the intensive interaction of those locals with other members of the team, both while collecting data and during the iterative data collection/data analysis process, provides opportunities for informal gossip.

The view that it takes extended periods of time to get information about social change assumes that only very long-term participant observation can get at changes over several decades. This assumption appears to consider only the length of a single, initial period of fieldwork and to ignore returns to the field to conduct follow-up studies. Not only is there nothing that precludes additional RQIs a decade after the first one, follow-up RQIs are probably more likely, since the critical resource limiting the return to the field by traditional researchers is often their lack of time for prolonged fieldwork.

There is some recognition that "traditions" are one source of the contention by some that extended time in the field is required for research. Wolcott commented in the 1999 edition of his book:

> To an old-time and old-fashioned ethnographer like me, terms like ethnography or fieldwork join uneasily with a qualifier like rapid. . . . My motto, to "do less, more thoroughly," may be nothing more than rationalization for my preferred and accustomed pace. Perhaps I envision a fieldwork entirely of my own making, having mistakenly accepted pronouncement about its duration (such as "one year at the least, and preferably two") as minimum standards when today's fieldworkers regard them as impractical and unnecessary. (1999, 110)

There may be some situations in which prolonged fieldwork is required to gain access and build rapport, in which situations are extremely complex, or a long cycle needs to be monitored. However, for most situations, RQI should produce a sufficiently rich understanding of the insider perspective for the design of additional research or to initiate activities that have to be started quickly.

Despite the strong case that prolonged fieldwork often is not required to achieve the goals of qualitative research, many researchers can be expected to remain skeptical of the rapidity of RQIs. As long as they conceive of RQI as a one-off exercise by outsiders who lack an initial understanding and familiarity with the environment, they are likely to view any information or knowledge gained from RQI as superficial (Leurs 1997, 292). My goal is to help anyone with these concerns understand the role of intensive teamwork in an RQI. As noted by Bernard (2011, 264), applied researchers do not "have the luxury" of doing long-term fieldwork and needs assessments often have to be done "in a few weeks."

> For many situations, RQI should provide enough of the insiders' perspectives to initiate activities or additional research.

RQI and Support for Long-Term Fieldwork

As long as people have been doing rapid research, others have been suggesting that the real danger is that the rapid work is being done instead of the long-term work that is needed. The implication sometimes seems to be that the rapid approaches are chosen because they are cheaper and that if they were not available, these resources could be used for long-term fieldwork. Some fear RQI will undermine support for less glamorous work (see Abate 1992, 486). The problem with this argument is that it assumes that there is sufficient time for implementing long-term work. In many situations where RQI is used, results are needed immediately to design interventions for problems that will not wait. Often it is not a question of doing rapid versus long-term fieldwork. The option of long-term fieldwork is not there. In some cases there might be time to do more prolonged fieldwork, but the resources are not there. Carrying out an RQI is not as cheap as some assume, but it is almost always cheaper than conventional, long-term work. Where time and resources are limited, the alternatives are either no inquiry (and research tourism is not counted as inquiry) or inappropriate research, such as questionnaire survey research without an understanding of the categories used by local people, because it is believed to be quick.

RQI can complement long-term research.

Rapid research can complement long-term research in situations where the time and funding are available. One of the goals of RQI is to identify when further research is needed. An RQI can identify situations where longer qualitative work is needed to establish rapport, deal with extremely complex issues, or monitor a long-term process like a growing season or language acquisition by a child. RQI can also provide the categories and terms to make questionnaire research meaningful in situations where categories and specific terms used by local people are not known. Many of the advocates of RQI agree with Emmons, a research associate at the Smithsonian Institution, that whenever possible rapid research is intended to supplement, not replace, long-term fieldwork (Abate 1992, 487).

Students who use RQI or a Mini-RAP to do short-term course assignments can be expected to take away from the experience appreciation for the strength of qualitative research for developing understanding of a situation while recognizing some of the problems with the most common alternatives. One lesson students need to learn is that the choice of an approach to research is not a moral issue, but one of matching needs with resources.

Concerns about RQI

The following concerns affect people's trust in the results of RQI or limit its usefulness. In some cases, there is not much that can be done about a concern, but in most cases knowing about it will allow the RQI team to address it.

Too Little or Too Much Time

Even though I am convinced qualitative inquiry can be done quickly, I am also convinced that there is a minimum time requirement. Two points should be emphasized. One, more time in the field will inevitably produce better results, with the possibility of some limited exceptions discussed below. Two, spending too little time on the RQI is the most serious threat to its rigor. Team members need sufficient time to be observant and have time for just talking and having a good conversation (Chambers 2008, 81). Attempts at RQI carried out with insufficient time and inadequate planning should probably be called "research tourism." One of my students suggested another name (one with an appropriate acronym) for RAP done too quickly: Condensed Rapid Assessment Process (CRAP).

Inadequate time introduces predictable biases into the process, including inappropriate focus on elements of the situation that are most obvious, observation of situations that are easiest to observe, contact with individuals already involved in change, and contact with individuals who are less disadvantaged (Chambers 1980, 3). Inadequate time can also result in not enough attention to the relationships, and may result in a failure to recognize that what is seen is a moment in time and is not necessarily the long-term trend. The length of an RQI will depend upon the situation, but anything less than the equivalent of four or five days is probably inadequate for carrying out discussions; identifying, discussing, modifying, and rejecting ideas that emerge from these discussions; and putting these ideas together in a usable form.

> Less than four or five days is probably inadequate!

There are some situations where results must be produced in a timely way if they are to be useful. Additional field time might improve the precision but would make the results useless. When resources are limited, the choice might be between having a lone researcher for a longer period or a team for a shorter period of time. In some cases shorter fieldwork by a team may be more useful than longer-term research by a lone researcher. Complex situations may require the expertise of several disciplines. There are also isolated situations in which additional time in the field communicates the wrong message about the confidence one can have in the results. Longer fieldwork may result in increased confidence in the results, when healthy skepticism is appropriate. Finally, an RQI that is too long may waste time and cause both insiders and outsiders to view the RQI as an end in itself, instead of as a tool for starting the learning process.

Cultural Appropriateness

The fundamental role of teams and the inclusion of local people on these teams may make RQI culturally inappropriate for some situations. Given the public nature of much RQI work, one should expect serious distortions in cultures where women do not participate openly in public life. Mixed-gender teams may be especially inappropriate for some topics and may have an inhibiting effect on the participation of some (Leurs 1997, 292). The RQI teams should be especially sensitive to the cultural appropriateness of using techniques like rich picture with people more comfortable with telling stories, rankings with people reluctant to compare each other, or audio recording in a wide range of situations.

Political and Economic Context

An accurate assessment of a situation should include a full description of the context. According to Manderson and Asby (1992 as cited in Harris, Jerome, and Fawcett 1997, 376), rapid research methods often consider the social and cultural context, but do not consider the political and economic context. Consideration of the political and economic context is closely tied to issues of differences in power and the relationship between the sponsors, the local stakeholders, and the RQI team. I will return to this issue in chapter 8.

Problems with Team Composition

For the results of RQI to be useful, the assessment team must be credible. Using multidisciplinary RQI teams generally increases credibility.

The lack of gender diversity on many teams continues to be a problem. Gender diversity may be vital in establishing the credibility and trustworthiness of the team with community members (Harris et al. 1997, 376).

The quality of RQI increases when team members have experience in their respective disciplines. According to Abate (1992) only when researchers "know how to do it right can they do it fast" (486). There is some concern that the shortage of support for long-term studies may tempt funding agencies to let less experienced researchers undertake rapid research before they have proven themselves in traditional scientific fieldwork (486). Even though RQI teams need members with both technical expertise in their discipline and an understanding of qualitative research, the success of RQI does not depend upon "superstars." I do not accept Macintyre's (1995) argument that a serious weakness of rapid research is its dependence for quality on the caliber of the expert team members. Assuming team members have an adequate level of technical expertise, they should be able to develop sufficient understanding of the principles of qualitative research to be able to select appropriate research techniques. Given the right attitude, with practice these skills should improve, especially if attention is given to identifying lessons from the implementation of each RQI.

Problems with Choice of Respondents and Informants

It is often easier to find people to talk with who are better off, more articulate, and more like the members of the RQI team. It is far more difficult to find the poorer, less articulate, and more upset, and those least like the members of the RQI team to engage with. The RAP Sheet has been designed to remind the team of the need for diversity in whom they talk to, including finding and talking to the "troublemakers." Chambers (2008) identified numerous biases that as of 2008 were being addressed to some extent but that still resulted in "poverty unperceived" including visiting projects, failure to ask about who is being left out, concerns about security, and not visiting or seeing the poor (41–45).

Inappropriate Use

Rapid research methods are not appropriate for some situations. Usually, rapid research methods should not be used for estimating numbers or percentages of a population with specific characteristics. It is therefore not very surprising that an Expanded Programme of Immunization study in Nigeria suggested an immunization rate of approximately 80 percent, while a demographic and health survey found a rate of only 50 percent for the same age children. A statistician commenting on the Nigerian results said that it was his opinion that "health managers are only fooling themselves if they rely on this methodology for estimates of coverage." Sampling procedures and the lack of trustworthy data on households to sample are some of the problems with the use of rapid research to estimate numbers (Macintyre 1995, 5).

Problems with Credibility

Many of the Rapid Assessment procedures were developed in response to calls from funding agencies for faster qualitative field research (Kumar 1987). While these methods have been recognized by many in the international development community as vital for in-depth understanding of problems, there are still some who are uncomfortable with them. One of the most frequently identified problems with rapid research methods is that they do not produce numbers. Usually, if numbers are needed, RQI is an inappropriate method, especially if it is to be used alone. Funding agencies need to recognize that there are many situations in which an understanding of a local situation using the terms and categories of the local participants is sufficient for the initial design of an intervention. There are other times when the results of RQI will be critical for the design of additional research or for monitoring the implementation of activities. Practitioners have a responsibility to resist pressure from funding organizations to use rapid methods in inappropriate situations. If rapid methods are used in inappropriate situations, the results will be subject to valid criticism.

Chambers (1997) lamented that calling research methods rapid has "been used to justify and legitimize sloppy, biased, rushed, and unself-critical work." Despite the large number of rapid research reports that have been done during the last several decades, there has been very little critical evaluation of methods by practitioners. This type of criticism is needed to improve the methodology and increase the credibility of the results. Chambers advised, "Embrace error. We all make mistakes, and do things badly sometimes. . . . Don't hide it. Share it." (See ch. 1, "The Need for Caution about the Use of RQI.")

Despite calls for verification of the results of rapid research such as RQI by the use of conventional research, there are very few examples where this has occurred. One example of verification compared the results of Rapid Assessment Methods (RAMs) of the health of a complex ecological wetland with traditional largely quantitative methods. Results from the different approaches to research were found to "correspond" and the authors concluded that "rapid assessments have the ability to transform the way we approach wetland evaluation and to more effectively inform management decisions" (Stein et al. 2009, 1–6).

Publication of the Results of Rapid Research

Despite the presence of four world-renowned scientists on one of the early rapid research studies by Conservation International, the credibility of their results was questioned because they did not publish their results in a peer-reviewed journal (Abate 1992, 486). The response of one of the team members, Ted Parker, was "Ten or 15 years from now our scientific contributions won't be important, but there may be some places that still exist because of what we've done" (486).

Peer review by colleagues of the results of RQI is desirable and is encouraged. Publishing results in journals is one, but not the only, way to achieve this. Table 6.1 contains a list of forty-one journals that published articles based on rapid research methods (see ch. 10, "Sharing RQI Reports").

Dealing with Flexibility

One of the major challenges for RQI is to sustain and enhance innovations and a willingness to experiment with new methods without losing rigor. Flexibility is critical to making RQI relevant to a wide range of systems and is a major strength of the approach. However, this flexibility can be abused and has been interpreted by some as allowing individuals to do anything, or almost nothing, and call it "RAP" or "RQI." A set of standard techniques could solve this problem, but only at the expense of the needed flexibility. One of the most common criticisms of rapid qualitative research methods has been their arbitrary "laundry-list type elements" (Anker 1991 cited in Macintyre 1995).

Table 6.1. Examples of Journals That Have Published Articles Based on Rapid Research, 1998–2012

Journals
African Journal of Ecology
Agricultural Systems
Archives of Physical Medicine and Rehabilitation
Australasian Journal of Early Childhood
BioScience
Bulletin of the American Meteorological Society
Cities and the Environment
Environment and Urbanization
Epidemiology and Social Science
Fish and Fisheries
Health Promotion Practice
High Plains Applied Anthropologist
International Information and Library Review
International Journal of Health Planning and Management
International Nursing Review
International Quarterly of Community Health Education
Internet Research
Journal of Adult Education
Journal of International Agricultural and Extension Education
Journal of Prevention and Intervention in the Community
Journal of Public Health
Journal of Rehabilitation Research
Journal of the Association of Nurses in AIDS Care
Landscape and Urban Planning
Maternal and Child Health Journal
Medical Informatics and Decision Making
Military Medicine
Nurse Educator
Nursing Inquiry
Permanente Journal
Practicing Anthropology
Qualitative Health Research
Research in the Schools
Reviews in Anthropology
Rural Society
Rural Sociology
Sexually Transmitted Diseases
Social Policy Journal of New Zealand
Social Science and Medicine
Studies in Health Technology and Informatics
Tropical Medicine and International Health

The RAP Sheet

The alternative to standardization is agreement on basic principles and then documentation, as part of the RQI report, of the specific techniques used. I propose the use of a checklist, the "RAP Sheet," to document what was done and to allow the reader to judge the quality of the work. The RAP Sheet also can remind the RQI team of important issues during the inquiry. The generic RAP Sheet in Figure 6.1 should be adapted to the specific situation of each RQI before it is used. I have concluded that RAP Sheet is a better name for the checklist than RQI Sheet because a "rap sheet" is usually associated with an individual's record with one meaning being what was determined during a trial before the "rap" of the judge's gavel marking the end of the trial. It is very unlikely

RAP Sheet

Title[1]:
Fieldwork dates:
Objectives:
RQI Team Members:

Name	Tech. Background	Language Use[2]	Local/ Outsider[3]	RQI Experience[4]

Hours in field collecting data:
Hours team discussing data:
Information collected in advance and reviewed by the team:
Examples of information collected by direct observation:
Number of participants interviewed:
Location of interviews:
Method of selection of respondents:
Among individual respondents, approximate percentage that were:
_____% women, _____% old people, _____% youth,
_____% among the poorest 25 percent, _____% living off road
_____% from significant ethnic or cultural minorities
_____% from those identified as troublemakers
_____% ______________________________
_____% ______________________________
Number of key informants interviewed:
Key informants positions/occupations:
Topics key informants reported on:
Topics for group interviews and composition of groups:
Date set for reviewing and updating this report:

---

[1] The title should include the name of the geographic or administrative unit.
[2] Language use categories:

A. Exclusive use of respondents' first language
B. Use of respondents' second language
C. Mixture of respondents' first and second languages
D. Mixture of respondents' languages and use of interpreter
E. Exclusive use of interpreters

[3] Insider or Outsider
[4] Prior experience

N No prior experience doing RQI
T Participation in a training course on RQI
1 to *n* Number of prior RQIs

Figure 6.1. Sample RAP Sheet

that all the items on the sample RAP Sheet will be relevant to any specific RQI, and even if they are it could make the checklist too long to be useful. The RAP Sheet is loosely based on the Human Relations Area Files data quality control schedule (see Lagacé 1970).

> Modify first and then use the RAP Sheet.

The Rap on Bad RQI and Good RQI

If done too quickly and without sufficient methodological rigor, RQI can be more dangerous than "research tourism." At a minimum there is a need for team orientation, and for several cycles of data collection and analysis. Ensuring data from multiple sources and triangulation requires a minimum amount of time for multiple semistructured interviews and for careful observations that are most relevant to the situation being investigated. An iterative approach to data collection and analysis requires time for the team to meet, to carefully consider what it has found, to develop and consider data displays, and to explore tentative conclusions before returning to the field for additional data collection. Checking back with the individuals who have provided information takes time, as does the preparation of a report.

When there is not enough time to do RQI rigorously, it is probably better to do "research tourism" and to clearly call it research tourism. Decisions based on poor data are no more likely to be good decisions than decisions based on no data at all (Macintyre 1995).

RQI provides relatively quick qualitative results that are likely to be sound enough that they can be used for decisions about additional research or preliminary decisions for the design and implementation of applied activities. When applied with care and caution, RQI can help prevent the errors that can result from "tourism" or the inappropriate use of a questionnaire survey before a local situation is sufficiently well understood to formulate questions. I am not suggesting that RQI can substitute for long-term, in-depth studies, where a situation calls for that and time and resources are available. I am suggesting that, in many situations, RQI will produce sufficiently solid results for the design of additional research or to initiate activities that have to be started promptly.

When RQI/Mini-RAP is an activity in a research course, students can be expected to take away from the activity a sophisticated understanding of some of the important issues concerning qualitative research. Issues of flexibility, time in the field, evidence, presentation, and trustworthiness are not limited to RQI.

Additional Readings

One of the central themes of this chapter has been that the credibility of RQI does not depend upon the use of specific techniques. Given the impossibility of developing a checklist for evaluating credibility, the issue becomes complicated. Key issues include, but are not limited to, the nature of qualitative research and appropriate use of qualitative research. Willis's (2007) chapter provides philosophical foundations for examining these issues and introduces the concepts of poetic logic and complexity theory. Denzin's (2011 and a slightly different version 2009, chapter 4) article on the politics of evidence focuses on both the assumptions and ethics involved with special attention to who determines the standards. Saldaña (2011) gives special attention to the issue of presentation as a credibility issue in his chapter 6 on "Writing and Presenting Qualitative Research." Finally, Shenton (2004) provides a short summary of some of the concepts related to trustworthiness. Utarini at al. (2001) identified eleven critical criteria for appraising studies based on rapid research methods. These criteria cover issues from preparation to presentation of findings. Utarini et al. do not call for the use of specific techniques but suggest greater attention to these issues would enhance the strength of rapid studies.

Denzin, Norman R. 2011. The politics of evidence. In *The Sage handbook of qualitative research*, edited by Norman K. Denzin and Yvonna S. Lincoln, 645–58. Los Angeles, CA: Sage.

———. 2009. *Qualitative inquiry under fire: Toward a new paradigm dialogue.* Walnut Creek, CA: West Coast Press.

Saldaña, Johnny. 2011. *Fundamentals of qualitative research: Understanding qualitative research.* New York: Oxford University Press.

Shenton, Andrew. 2004. Strategies of ensuring trustworthiness in qualitative research project. *Education for Information* 22: 63–75.

Utarini, Adi, Anna Winkvist, and Gretel H. Pelto. 2001. Appraising studies in health using rapid assessment procedures (RAP): Eleven critical criteria. *Human Organization* 60(4): 390–400.

Willis, Jerry, with Muktha Jost, and Rema Nilakanta. 2007. *Foundations of qualitative research: Interpretive and critical approaches.* Thousand Oaks, CA: Sage.

CHAPTER SEVEN
TECHNOLOGY FOR IMPROVING AND SPEEDING UP RQI

Lauren Angelone

Main Points

1. Instead of creating distance between researchers and understanding of a situation, technology can facilitate a team approach to understanding the insider's perspective.
2. Technology can make an enormous amount of rich data accessible to the entire RQI team.
3. Documents can be crafted collaboratively by team members and with participants and can become pieces of collective intelligence.
4. Social media can enable everyone involved in the RQI process to be creators and contributors.
5. Documents may need to be encrypted before saving them to the cloud.
6. Technology can aid the RQI team in collaboration. Tools allow RQI team members to store, synchronize, and organize materials so they are accessible to all members of the team.
7. Technology tools can assist the RQI team in collecting and triangulating data. Specific tools allow the team to take audio and video notes and capture images, and improve the interview process.
8. Technology tools can facilitate interactive data analysis. Especially relevant are manual and voice recognition transcribing tools and tools for coding and searching for themes.
9. There are several websites that provide information on the most current technology tools for qualitative research.

Figure 7.1. Main Points as a Word Cloud

Technology and Qualitative Research

Technology, in various forms, has always been present in qualitative inquiry. Most of these tools, such as pen and paper, word processors, tape recorders, and so on, are considered low tech and the field of social science has sometimes argued for the "intimacy" with data collection and analysis that is supported by these "raw" options. But technology does not necessarily have to distance researchers from the version of reality under study. Davidson and di Gregorio (2011, 627) even suggested technology offers qualitative researchers the following advantages: (a) a digital location for organizing all the materials related to a study, (b) tools that can be applied to organize these materials, (c) portability, and (d) transparency for the researcher and others to view and reflect on the materials. Since 2000 the technology available to researchers has expanded and increased options that can support and enhance data collection and analysis for qualitative research that is collaborative, creative, and more rapid.

> Technology can provide (a) a digital location for materials, (b) help organizing the materials, (c) a way to make material portable and sharable, and (d) increased transparency and the ability to reflect on material.

The increases in capabilities of both hardware and software have implications for qualitative inquiry. Hard drives have a larger storage capacity and are less expensive than ever before. In addition, storage on third-party servers, also referred to as "the cloud," is readily available. This allows researchers to store an enormous amount of rich data, field notes, and transcripts, but also multimedia including digital recordings of interviews, photos, and video. These digital recordings are also easier to take using personal devices like smartphones that now have the capability to function as high-quality microphones, cameras, and video cameras. Compact cameras/video cameras are also relatively inexpensive, high quality, and able to hold large amounts of data.

This chapter will focus on the types of software and hardware available to aid qualitative researchers in collecting, sharing, and analyzing data. This chapter discusses technology that was available in 2014 and evaluations should always be read "as of the time of the preparation of this chapter."

As a first step, switching from handwritten notes to word-processed notes allows them to be more easily shared, duplicated, and stored in multiple locations, but also searched and run through different analysis programs to help the researcher see the data in new ways. As a second step, if these notes are typed or copied into collaborative writing programs (such as Google Docs) rather than static ones (Microsoft Word, for example), the notes not only become sharable, but documents can be crafted collectively with other research team members or participants.

Word-processed notes can be shared, duplicated, stored, and searched.

Collaborative writing programs are tools that have emerged from what has been referred to as web 2.0 or the participatory web (Greenhow, Robelia, and Hughes 2009, 247). Web 2.0 is the second iteration of the Internet. In the first iteration, the Internet was mostly for one-way information with very few users being able to contribute to or interact with that information. In web 2.0, users participate and create together. Social media, like Facebook and Twitter, are good examples where all users are creators and contributors rather than passive consumers.

Web 2.0 can be beneficial to qualitative researchers in several ways. In addition to working collaboratively with other research team members, research teams, whether in the United States or overseas, can work collaboratively online with their colleagues back home or with participants. Documents can become pieces of collective intelligence (Lévy 1997, 2001). There are also many free tools beyond collaborative writing programs that can be beneficial for collaboration, project organization, and creativity. These tools can facilitate the creation of mindmaps, interactive multimodal maps, wikis, and photo sharing, as well as audio and video transcription. Reflexivity, or the act of reflecting on one's own thinking in the context of social theory and research, can be a collective process facilitated by use of tools like blogs that allow multiple users to comment and contribute in a journal-like fashion. I will explore several examples of these types of tools in this chapter.

Expect all technology to change including backup tools.

There are several drawbacks to using technology that should be taken into consideration by the RQI team. Recording digital material requires making several backups. One backup should be on a third-party server, so that no matter what happens to devices, the data will not be lost. Use of third-party servers though, depending on the sensitivity of the research, raises questions about who owns the information and how secure it is. A paid service may be more appropriate for data than a free one and encryption may be needed. Also, if the team decides to take advantage of free and collaborative tools available via the web, they must check the Internet connections in the field that they will be visiting. And if they are traveling overseas, they should also check the availability of some of the specific programs they plan to use. Another problem with becoming accustomed to many of these tools is that they change frequently. Updates are commonly being made, but more than that, companies go out of business and some tools can disappear quickly. For this reason, it is probably best to go with well-established tools.

Intent of the Chapter

Tools identified in the text are a sample and are meant to provide awareness of what was available when this review was prepared.

The purpose of the rest of this chapter is to introduce digital tools that can support, enhance, and speed up team-based qualitative inquiry. An overview of each tool and an example of how these tools could be used in RQI will be given without focusing on specific step-by-step instructions. There are several alternatives for most of the tools listed here and some of the unlisted technology may be a better fit. As such, this chapter is meant to make the reader aware of what is available as of the time this material was written and how it could be useful, rather than getting into detail that will quickly become out of date and insufficient. Tools will be put into categories that coincide with the RQI approach: enhanced teamwork; collaboration for data collection and triangulation; and tools for iterative data analysis. The assumption is that all of these tools can facilitate getting the insider's perspective and that there is not one set of tools that are especially relevant to this objective. Within these categories, a few well-known

As members of the RQI team begin to experiment and learn some of the programs listed in this chapter, as well as ones that they may find on their own, they need to keep in mind that there can be a steep learning curve that can be frustrating, but that the payoff can be worth the effort.

1. Technology can take time, especially when learning or setting up something new. This seems counterintuitive because technology is supposed to be the ultimate time-saver. But, if one can come to terms with the fact that it takes time to save time, she will save a lot of stress. Rest assured that given enough time, things will eventually work, but it may not be as simple or quick as anticipated.

2. One does not have to understand everything to figure something out. Learning how to use new tools on a computer can be daunting, but researchers do not have to be a computer scientist to be able to try new things and even get good at them. Oftentimes things seem more complicated than they really are. When all else fails try a Google search. There are forums with people having the same problem as well as manuals, tutorials, YouTube videos, etc. Online sources can be the greatest resource for ideas and explanations for experimenting with technology.

4. My best advice: Don't be afraid to make mistakes. It is not necessarily true that children are so much better at using the computer; it is that they are unafraid to make mistakes. Be like a kid and dive right in. If you make a mistake, it will not ruin everything and even if it does, lessons can be learned from mistakes. This is particularly true when experimenting with technology.

5. There are multiple ways to do the same thing. When using a computer, most users soon find out, if they did not already know, that there is not one right way to do anything. It is not necessary to memorize how to complete different tasks. Once some skills are acquired, mostly through trial and error, these skills can be applied to new programs/environments. People often do the same thing in different ways depending on their comfort level. Be open to learning new and possibly easier ways to do things.

6. Technology is constantly evolving and changing. This is important to learn because few can keep up with the pace of change and even fewer can be aware of the breadth of information and tools available on the Internet. Books are published such as this one and are immediately out of date. This is the nature of technology. The rapid rate of change makes it imperative that users collaborate and share emerging or little known technologies as they find them.

7. Free is good, but sometimes it is worth it to shell out some money. There are many excellent resources free on the Internet. With the growing popularity of open-source programs, free quality programs are more easily accessible than ever. However, depending on what you plan to use the technology for, you may need to purchase a program or, more often, equipment that is necessary for the job. Most of these things can be borrowed or found in a lab, but sooner or later, it is better to own quality items. Sometimes used equipment can be sold for a decent price to minimize the cost of upgrading when it eventually becomes necessary.

Figure 7.2. "Rules" of Technology Use to Avoid Utter Frustration

examples of available technology tools will be introduced and discussed as they relate to RQI specifically.

Table 7.1 shows the relationship of selected, illustrative technology tools to team-based RQI. The left column lists tools that subjectively were considered the leading tools as of 2014 while the right column lists alternative or additional tools.

Table 7.1. Relationship of Illustrative Technology Tools to Team-Based RQI

Primary Technology	*Other Technologies*
I. Enhanced teamwork and collaboration	
A. Project organization	
Dropbox	Windows Live Mesh SpiderOak SugarSync Wuala
B. Collaborative writing	
Google Drive and Suite of Programs from Google	Zoho Docs Office Web Apps Evernote
C. Mindmapping (diagram used to visually outline information)	
Coggle is a free web application	Mindmeister Bubbl.us Mind42
D. Communication-at-a-distance online video conferencing	
Skype	FaceTime Google Hangouts
E. Blogs	
Blogger part of the Google suite	Wordpress
II. Collaboration for Data Collection and Triangulation	
A. Observation tools, audio and video notes, and image capture	
Evernote	Google Keep Google Docs Microsoft OneNote Awesome Note
B Photo hosting, sharing, organizing	
Flickr	Picasa PhotoBucket SmugMug Evernote
C. Interviews digital voice recording	
Smartphones Audacity	Laptops Digital recorders Microphone attachments iRig MIC Cast
D. Video recording	
Sony Bloggie camera	Smartphone but limited storage

(continued)

Table 7.1. *(Continued)*

Primary Technology	*Other Technologies*
III. Iterative Data Analysis	
A. Transcribing voice recognition	
Dragon Naturally Speaking	
B. Transcribing manually	
Express Scribe	f4
	optional foot pedal
C. Transcribing video	
Transana	f4
D. Coding/Themes, Computer-Assisted Qualitative Data Analysis Software (CAQDAS)	
NVivo	ATLAS.ti
	MAXQDA
	HyperRESEARCH
	Coding Analysis Toolkit (CAT)
	QDA Miner Lite
	TAMS Analyzer
E. Word clouds	
Wordle	TagCrowd
	WordSift
	TAGUL

Enhanced Teamwork

The first category of tools that will be covered are tools that can aid the RQI team in collaboration. The main function of web 2.0 is communication, so it follows that researchers can make use of this attribute to help project organization, communication at a distance, and group reflexivity. This is particularly beneficial to researchers utilizing RQI as it streamlines the collaborative efforts of the team as they work to begin the process of understanding complex relationships in a short amount of time with a focus on the insider's perspective.

There are numerous programs that allow RQI team members to store, synchronize (update documents locally and on the cloud), and organize documents so they are accessible to all members of the team. This allows the RQI team to avoid passing around a flash drive or e-mailing documents back and forth, a process that makes it difficult to determine which document is the most recent. This also provides a backup of all of the data in a secure location. If one uses a program like this, and keeps a copy of the data on the computer's hard drive, that folder needs to be synced with the folder stored in the cloud. This means that if the computer is stolen or destroyed, the RQI team will still have access to their documents. This synced folder can also be accessed by members of the team from a variety of devices, including a laptop, tablet, or smartphone, provided that these devices have access to the Internet. Some smartphones have an application (app) to make accessing documents easier.

> Tools for team collaboration, such as Dropbox, are especially relevant to team-based RQI.

Dropbox (http://dropbox.com) is a good example of a program that allows the RQI team to store, sync, and organize data in the cloud. Dropbox provides 2GB of storage for free, but requires payment beyond that. When one signs up for Dropbox, the user designates a folder on the local hard drive that will be synced in the cloud. Several other folders can be placed within this folder, but designating a single folder allows Dropbox to automatically save any changes that you make as you write and save documents to the cloud. That means that it is not necessary to upload new versions of single documents over and over again as must be done on a flash drive or external hard drive or if documents are e-mailed to team members. To ensure the security of data, I recommend that you encrypt any sensitive documents or folders prior to saving them into your Dropbox folder. Encryption can be as simple as right clicking on a folder on computers running a Windows operating system. Programs similar to Dropbox include Microsoft's OneDrive (https://onedrive.live.com/about/en-us/), SpiderOak (https://spideroak.com/), SugarSync (https://www.sugarsync.com/), and Wuala (http://www.wuala.com/).

> Collaborative writing programs, such as Google Drive, allow team members to co-create understanding.

Collaborative writing programs are also tools that permit users to store and organize data in the cloud. The main function of these programs is to allow users to write together synchronously or asynchronously.

This functionality will be explored further in the triangulation portion of this chapter, but these programs also save documents to the cloud as one writes, as long as Internet connection is available, without requiring a special folder on the computer or syncing. The downside to collaborative writing programs is that there are no local copies of the documents stored on the local computer if the Internet is not available or the servers are down. The user can choose to download copies to their computer, but these copies would quickly become out of date as changes are made to the documents in the cloud. Collaborative writing programs also allow groups of users to collectively create folders to organize documents.

Google has a suite of tools and server space that are free to use if one has signed up for a Gmail account. As of 2014, Google Drive (https://drive.google.com) provided 15GB of space for users to write, save, and organize documents and photos individually or collaboratively. Users can upload documents or create them within Google Drive and then edit them and they will be automatically saved. When collaborators log in, they will only be able to access the most updated version of the document and file organization. Like Dropbox, Google Drive is hosted on a third-party server so breeches of privacy, though rare, are possible. Google suggests enrolling in their two-step verification process that requires a password and a code sent to your phone. In 2014 Google Drive was the standard in this type of storage space and collaborative writing applications, but similar programs include Zoho Docs (https://www.zoho.com/docs/), Office Web Apps (http://office.microsoft.com/en-us/web-apps/), and Evernote (http://evernote.com/). Evernote has more functions that will be explored further in the triangulation portion of this chapter.

Mindmapping tools are widely available on the Internet and can help researchers as they collaborate on the organization of their research project. Mindmapping can also be used for brainstorming and preliminary data analysis and these functions will be explored further in the section on data analysis. Using a mindmap for organization throughout fieldwork can help qualitative researchers see how parts of the research process connect and what parts may be missing. Web 2.0 mindmaps have several advantages over mindmaps on chart paper or even mindmaps that require a program to be downloaded to your hard drive. They are easily modified and accessible from any Internet-connected device and can be edited by multiple users.

Coggle (http://coggle.it/) is a free web application, with nothing to download to the local computer, which allows users to collaboratively create mindmaps. Mindmaps can be customized by type of map and color of connection. Photos can also be dragged into the program for a multimodal map. Coggle allows users to share their mindmaps with other users who may also edit the mindmap. Coggle mindmaps can be downloaded to a hard drive as a PDF or PNG image. Other web-based mindmaps tools include mindmeister (http://www.mindmeister.com/), bubbl.us (https://bubbl.us/), and Mind42 (http://mind42.com/). These tools are either completely free or there is a free introductory version.

Team Communication at a Distance

Online video conferencing tools allow users to video chat over the Internet. Video conferencing is an enhancement to other forms of communication at a distance as these tools add video to typical phone conversations and do not require a landline. They can also give the feeling of a more intimate type of communication than e-mail or phone conversations, something that may benefit the RQI researcher who is communicating at a distance with either fellow RQI team members or participants. When speaking with participants in particular, it may be beneficial to see their body language and expressions as they speak. These conversations also have the potential to be recorded and reviewed using a screen recording software such as Evaer (http://www.evaer.com/) for Windows and eCamm (http://www.ecamm.com/mac/callrecorder/) for Mac. The main drawback to video conferencing tools is that they rely on a robust Internet connection and delays are common.

Skype (http://www.skype.com) is the most established of the online video conferencing tools. Its basic capabilities are free and simple to use. There is typically a short delay on even the fastest Internet connection that can be disruptive to the flow of conversation, but the quality is good compared to other comparable tools. In order to make free video calls, both users must have a Skype account, a web cam, and Internet access. Alternatives to Skype include FaceTime (http://www.apple.com/mac/facetime), which makes video calling easier on mobile devices, and Google Hangouts (http://www.google.com/+/learnmore/hangouts), which as of 2014 allow for free group video chats of up to nine people.

Facilitating Reflexivity

Blogs can facilitate team reflexivity. Short for weblogs, blogs are essentially online journals. Blogs have been used as personal journals, but also as websites for keeping up with changing conditions such as politics or events impacting an RQI. Blogs place the newest post first and it becomes the first thing that a visitor to the site reads. Blogs are meant to be dynamic websites, constantly updated with the latest information. Blogs can be a convenient tool for the reflexive researcher or team. Since a blog is a website, it can be accessed and edited from any Internet-connected device. Blogs also allow multiple users to post and for visitors to the site to comment. This would allow a team of researchers to not only share their thoughts, but to have a discussion around those thoughts. And, if necessary, blogs can be made private to protect sensitive data. Like many of the tools listed however, blogs are typically hosted by a third-party, so efforts should be made to anonymize data.

Blogger (http://www.blogger.com) is a part of the Google suite of tools that one can access when signing up for a Gmail account. Blogger is an intuitive blogging platform that is good for beginning bloggers. Posts can be made simply with one click and customization, though limited, is available as well. Wordpress (http://www.wordpress.com) is another blogging platform. Wordpress is also relatively easy to use for a beginner, but Wordpress tends to have a more professional look and there are many more options available for customization of the site. Both of these blogging platforms are free and allow multiple users to blog together and visitors to the site to make comments. Either Blogger or Wordpress would make good collaborative blogs for RQI team reflexivity. (See Siltanen et al. 2008 for a discussion of reflexivity and teamwork.)

Collaboration for Data Collection and Triangulation

The second category of tools are those that can assist the RQI team in collecting and triangulating data. Some of these tools overlap with those discussed in the previous section, but rather than focusing on how they can enhance teamwork, their uses for gathering data and sharing those data specifically will be covered. In this section, researchers will learn ways to complete data collection and share data more efficiently.

Observation

There are several advanced note-taking tools that allow users to take audio and video notes and capture images and websites, in addition to the standard word-processed notes. These notes can also be tagged with keywords and are searchable, making it easier to search quickly for key phrases. Most of these note-taking tools allow users to share folders with other users and it can be useful for the RQI team to have access to all notes taken by all members of the team. These shared folders can also be accessed via multiple Internet-connected devices, such as a laptop, smartphone, or tablet, and notes added by any of these means can be automatically synced so that the research team is always seeing the latest version of each document.

As of 2014, Evernote was the standard note-taking tool. Evernote works on both Windows and Macintosh operating systems. It can be downloaded as a program to your computer or it can be used within a web browser. There are also apps to use and enhance Evernote on smartphones or tablets. For example, there is an app to integrate annotated maps with Evernote. Evernote allows folders to be shared with colleagues and made public if sensitive data is not an issue, but it also allows individual documents to be encrypted if necessary. Alternatives to Evernote include Google Keep (https://drive.google.com/keep/), Microsoft OneNote (http://office.microsoft.com/en-us/onenote/), and Awesome Note (http://www.bridworks.com/anote/eng/).

Google Docs could also be used as a collaborative note-taking tool. Google Docs allows users to write collaboratively and synchronously. This means that one can watch as a team member writes notes in the same document that she is working in, but from a different computer. This allows a synchronously collaborative experience, unlike Evernote in which documents are saved and then are available to other team members. Seeing changes being made by other RQI team members as they are being made could provide insight into their thoughts. It is important that team members take care not to get distracted while taking field notes by this or any other technology and lose focus on the goings-on in the field. In addition, having more than a few individuals writing together at the same time in Google Docs can cause confusion, as the sentences on the screen can appear to move up and down as others type.

When making observations in the field, the RQI team may also want to take photos. In addition to the record of events that photos provide, they also allow for a record of space relationship. There are several tools useful for collecting, organizing, and annotating photos. These tools can be used by taking photos on a smartphone and then uploading and manipulating photos through an app. Alternatively, photos can be taken with any digital camera and then uploaded through the website on a laptop or desktop. Either way, users are able to put photos into different albums, tag them with keywords, attach notes to them, and attach location tags. These albums can be public or private.

Flickr (https://www.flickr.com) is a well-established photo hosting and sharing site created by Yahoo! As of 2014, Flickr allowed 1 terabyte of storage for free, but users will see ads while browsing their photos. For an additional fee, users can have an ad-free account or more storage space. Flickr allows users to annotate photos in several ways. Notes can be added and users will see them as they roll over the image. Comments can also be made to the image and images can be placed on maps through the use of geotagging, which imprints images with location information. Albums of photos can be public or made private and shared with selected users. Flickr can also be integrated with Evernote or Google Drive using third-party sites. Alternatives to Flickr include Picasa (http://picasa.google.com/) that integrates with Google+, PhotoBucket (http://photobucket.com/), and SmugMug (http://www.smugmug.com/).

The ability to organize photos is also available within Evernote. Getting photos into Evernote is possible via the Evernote app and a smartphone camera, a laptop/desktop, and a webcam or by uploading photos from any digital camera to the Evernote website. Once in Evernote, photos can be grouped into folders and tagged, and notes can be added. In addition, Evernote has a special feature that can "read" text within images. For example, if you take an image of a stop sign and then search for the word *stop* in Evernote, the image of the stop sign will appear with a box around the word *stop*. The ability to identify text in a photo could be very useful to researchers that take numerous photos and are searching for a specific image. This feature could also be useful for early data analysis when searching for mentions of a specific word. The RQI researcher might consider using the ability to identify text in an image for document analysis. Evernote essentially digitizes documents to allow searching, so if a researcher takes a picture of a document, specific words can be searched for within Evernote.

Interviews

There are several ways to enhance, facilitate, and expedite the interview experience using digital tools. To begin with, if the researcher wants to avoid purchasing a handheld digital voice recorder with a singular purpose, smartphones are almost ubiquitously equipped with a built-in digital voice recorder. Though simple, these recorders are typically of decent quality and the files are able to be exported into file formats that can then be imported into transcription programs. There are, however, microphone attachments to smartphones that enhance the built-in smartphone microphone. For example, the iRig MIC Cast (http://www.ikmultimedia.com/products/irigmiccast/) is a compact microphone attachment that enhances the quality of voice recordings. This particular microphone also allows headphones to be plugged into it so that the researcher can monitor the sound quality throughout the interview. The iRig MIC Cast can be used with iOS (Apple) and Android devices.

Laptops generally also have built-in microphones. In order to record interviews quickly and inexpensively, the researcher could simply open up her laptop and record the interview using a free program like Audacity (http://audacity.sourceforge.net/). Audacity is a well-established program and a free download. It also makes it possible to edit audio recordings to eliminate background noise or to delete large portions of unnecessary recording. Audacity also allows for files to be easily exported in common file formats that can then be used in transcription programs.

If video interviews are preferable, smartphones are also equipped with built-in video recorders that are of increasingly good quality. These are convenient for short interviews and can be exported in file formats that allow for editing and analysis of the video. Smartphones, however, have limited storage space for large amounts of video since most smartphones have apps, photos, and other information that all take up space on the hard drive. It may be more practical to purchase a compact video recording device with a large built-in hard drive, such as a Sony Bloggie camera or something similar. The Sony Bloggie can record up to four hours of video that can then be exported onto a computer's hard drive and the Bloggie hard drive can be erased for more recording.

CHAPTER SEVEN

Tools for Iterative Data Analysis

The third category of tools that can support the work of the RQI team are tools for iterative data analysis. These tools can help in various stages of data analysis including transcribing audio and video interviews as well as in helping researchers organize data by coding. Word clouds are also useful for preliminary data analysis and will be explored further in this section.

Transcribing

Transcription tools facilitate transcription of audio or video files. They essentially combine audio or video files with a word-processing program. Some programs use voice recognition technology while other programs allow someone to listen to the recording and type what is heard. There are good voice recognition programs that can transcribe from audio files. As of 2014 the best-rated voice recognition software was Dragon Speech Recognition software, also called Dragon Naturally Speaking (http://www.nuance.com/dragon/index.htm). Dragon, like other voice recognition programs, has to be "taught" each individual voice and can learn only one voice. Since interviews involve more than one speaker, and RQI group interviews involve multiple team members as well as a respondent, voice recognition software will not work on a regular recording. The work-around to this problem is to download the recording of the raw interview that is to be transcribed to a computer. Then listen to the recording of the interview through headphones while activating the computer microphone with the voice recognition software running. Repeating the recorded text in a voice that has been taught to the software enables the software to transcribe the audio. The voice is turned into text as quickly as the person doing the transcribing can repeat clearly what is heard. The use of software for playing the raw interview that can be slowed down can improve the process and can minimize the need to pause or rewind the raw interview. There are descriptions of this process on the Internet. Applications for smartphones are available that work with the Dragon Speech Recognition Software (http://www.nuance.com/dragonmobile apps/).

> Voice recognition software, such as Dragon Naturally Speaking, can transcribe only one speaker. To transcribe an interview with two or more speakers requires a work-around such as after the interview one person listening to the recording and repeating what is heard.

Transcription tools simply provide a way to manually transcribe audio or video files. Express Scribe (http://www.nch.com.au/scribe/) is an easy-to-use transcription program. The basic program can be downloaded free of charge. Express Scribe allows users to import audio or video files and transcribe within the program. Users can set up hot keys to make starting and stopping the recording easier. It is also possible to play back recordings at slower or faster speeds. Users can then save the final transcribed document in a variety of formats, making it easier to import the text into other programs for analysis. A good alternative to Express Scribe is f4 (http://www.audiotranskription.de/english), which also allows for some coding and analysis. Both programs offer a foot pedal as an option for starting and stopping recordings hands free.

While Express Scribe supports the transcription of audio, Transana (http://www.transana.org/index.htm) is a program created specifically for transcribing video. Like f4, it also allows for some coding and analysis and will be discussed further in the next section on Computer-Assisted Qualitative Data Analysis Software. Transana allows users to import video and transcribe the video in sync with the text of the transcription. Unlike programs like Express Scribe, where text and audio/video are not linked, if you search for a specific phrase in Transana, you will find the video clip that matches that portion of the transcription. This is very helpful in situations where the space in which participants move is as important as the words that they say, such as during observations of a classroom environment or a visit to a farmer's field or a community center. The researcher not only can quickly find what was said, but she can also easily rewatch the context. In programs like Transana, clips can then be sorted into folders based on codes. This then allows the researcher to rewatch all clips related to a particular code without having to search through lengthy video; f4 also works in a similar fashion. These tools can be very helpful to an RQI team that has made the methodological decision to analyze video interviews and observations.

Coding and Identifying Themes

Computer-Assisted Qualitative Data Analysis Software (CAQDAS) are programs that support the analysis of qualitative data in various ways. Most programs allow the user to import text documents that can then be coded within the program by the RQI team. Users can code the textual data by selecting portions of the text and attaching one or more codes to it. The software can then run analyses based on searches, for example, for all text assigned an individual code or correlations between codes. More advanced programs can also code video and audio, and they can run more sophisticated analyses. Some programs can "learn" codes and apply them on their own. This has obvious dangers and there can be a loss of intimacy with the data. Researchers should always be sure to check that the codes being applied to text are appropriate.

> Computer-Assisted Qualitative Data Analysis Software (CAQDAS), such as NVivo, allows the RQI team to code transcripts of interviews and to run sophisticated queries on the data. However, RQI projects many not need or be able to justify the cost of programs such as NVivo but may find other software with more limited capabilities useful.

NVivo (http://www.qsrinternational.com/products_nvivo.aspx) is a well-known and well-established proprietary CAQDAS program. In 2014, it was in its tenth iteration. NVivo allows users to gather text, audio, video, website, and social media data and then run sophisticated queries on that data. These queries may allow the RQI team to see connections where they may not see them if handling the data manually. Though NVivo is not simple to learn, there are training tutorials and classes offered by the manufacturer and, once learned, NVivo can be a powerful resource for collecting, organizing, and analyzing data (see Bazeley and Jackson 2013). There are several alternatives to NVivo at various price points with different levels of sophistication. As of 2014, some of the most popular were ATLAS.ti (http://www.atlasti.com/), MAXQDA (http://www.maxqda.com/), and HyperRESEARCH (http://www.researchware.com/products/hyperresearch.html). When researching software options, the RQI team needs to be certain that the software is compatible with their operating system.

Though NVivo and other proprietary CAQDAS programs are useful for large projects and sophisticated queries, their hefty price tags can be prohibitive. There are several free and open-source programs available that may not have the extensive capabilities of programs like NVivo, but are certainly practical for many small-scale projects. Other free or open-source programs are Coding Analysis Toolkit, or CAT (http://cat.ucsur.pitt.edu/), QDA Miner Lite (http://provalisresearch.com/products/qualitative-data-analysis-software/freeware/), and TAMS Analyzer (http://tamsys.sourceforge.net/). None of these alternatives to NVivo are as intuitive or user friendly. CAT's advantage is that it is web based, so there is no program for the researcher to install on her computer. TAMS Analyzer is available for the Mac operating systems. QDA Miner Lite is based on the widely used QDA Miner.

While CAQDAS programs allow for high-level coding and analysis of data, word clouds are relatively new tools that, though simple, can also be useful to a researcher as they begin to look for codes and themes in their data. Word cloud generators take a block of text and search for the most frequently used words. They then create a cloud of words in which the largest words are the most frequently used. An entire interview transcript could be copied and pasted into a word cloud generator for a quick idea of what topics were discussed the most and in what ways. The use of word clouds for data analysis is limited, but useful.

Word clouds are simple, easy-to-use tools for identifying codes and themes.

Wordle (http://www.wordle.net/) is a well-established word cloud generator. It allows users to either copy and paste text or use the URL from a website with an RSS feed. Wordle allows users to customize the word cloud randomly or by changing the color, font, and layout of the cloud. Wordle also offers the option of viewing the number of times each word was used while filtering out common words like *the*. This chapter opens with a sample word cloud of the contents of this chapter created in Wordle in addition to a list of Main Points. Alternatives to Wordle are TagCrowd (http://tagcrowd.com/), WordSift (http://www.wordsift.com/), and TAGUL (http://tagul.com/).

CHAPTER SEVEN

Additional Reading and Other Resources

Digital tools for qualitative research are updated but perhaps not entirely new ways to support the organization, collection, and analysis of data. These tools can maximize efficiency by speeding up the process, but they can also allow the RQI team to see the data from different angles. Digital tools, as with any tools, should be carefully chosen by the RQI team based on the needs of the inquiry and should be carefully evaluated in terms of sensitivity of data and the availability of the Internet. Digital tools change quickly and frequently, so care must be taken to assure that the availability and quality of tools are maintained throughout the study. Because of the changes in these tools, it is difficult to write a chapter in a static analog object, such as this book, about digital tools for qualitative research. As such, I would like to direct you to the Internet for the most current information about technology tools for qualitative research and especially to Bamboo DiRT (http://dirt.projectbamboo.org/), a comprehensive resource of digital tools related to all types of scholarly work; the Mobile and Cloud Qualitative Research Apps, published in the *Qualitative Report*, a weekly online journal by Nova Southeastern University (http://www.nova.edu/ssss/QR/apps.html); and the Student Resources Site of *Digital Tools for Qualitative Research* (http://www.sagepub.com/paulus/study/default.htm).

The Mobile Reporting Field Guide prepared by students in the University of California, Berkeley, School of Journalism explores and evaluates smartphone tools for journalists. Many of these tools would also be relevant to qualitative researchers and RQI teams (http://multimediashooter.com/mobile/MobileGuide.pdf).

Two especially useful published resources are *Digital Tools for Qualitative Research* by Trena Paulus, Jessica Nina Lester, and Paul Dempster and "Qualitative Research and Technology: In the Midst of a Revolution" by Judith Davidson and Silvana di Gregorio.

Davidson, Judith, and Silvana di Gregorio. 2011. Qualitative research and technology: In the midst of a revolution. In *The Sage handbook of qualitative research*, 4th ed., edited by Norman K. Denzin and Yvonna S. Lincoln, 627–43. Thousand Oaks, CA: Sage.

Paulus, Trena, Jessica Nina Lester, and Paul Dempster. 2014. *Digital tools for qualitative research*. Thousand Oaks, CA: Sage.

CHAPTER EIGHT
WHO BENEFITS, WHO PAYS, AND WHO CALLS THE TUNE

Main Points

1. An RQI involves (1) the RQI team, (2) the stakeholders, and (3) the sponsors; interaction between these groups is influenced by differences in power and self-interests.
2. Even when the different groups are acting in good faith with each other, there can be problems.
3. Unequal power relationships both within the RQI team and between the RQI team and the local stakeholders can seriously threaten the process. Unequal power between the different groups involved in an RQI can limit the independence of the RQI team.
4. Researchers often promise to deliver more than is possible and sponsors often demand more than is reasonable.
5. RQI can help define issues from the perspective of local participants, including identifying special terms and definitions, but limited time precludes the depth of understanding that long-term fieldwork can produce.
6. Propriety refers to using procedures that are ethical and fair to those involved and affected.
7. Getting informed consent is essential for fair treatment.
8. The time people spend responding to the questions of an RQI team is valuable and this should be recognized by the RQI team.

9. Researchers rarely set out to betray the people who are the subjects of their inquiry, yet local participants are sometimes betrayed and often feel betrayed.
10. Betrayal is most likely to occur when the research team promises a level of confidentiality that is not delivered and when the research discloses information that should not be made public.
11. Bogus empowerment suggests to participants in research activities that their input will be acted upon when there is no intention of doing so. It promises but does not deliver the power people need to act on their own judgment.
12. Authentic empowerment gives people control over outcomes so they can be responsible for their activities.
13. The Institutional Review Board (IRB) process is critical for ensuring ethical treatment of participants and needs careful attention.
14. The Statement of Professional and Ethical Responsibilities of the Society for Applied Anthropology provides a summary of ethical issues that should be considered.
15. Completing an RQI/Mini-RAP as part of a course assignment can introduce ethical issues to students.

Ethics and RQI

The issues of benefits, costs, and control discussed in this chapter are not unique to RQI, but rather are issues for all research. However, these issues are especially relevant to qualitative research and even more so for applied qualitative research like RQI. Students who conduct short-term research such as a Mini-RAP as part of their course requirements need to be sensitive to ethical issues. Attention to the issues related to ethics and RQI, with special attention to ensuring informed consent, is as important as learning skills for doing qualitative research.

Unequal Power and Unequal Influence

An RQI involves three groups of players: (1) the RQI team, (2) the stakeholders, and (3) the sponsors. For an RQI, the stakeholders can be thought of as the individuals in the local group who are most likely to be impacted by the results of the RQI or any activities resulting from the

RQI. Stakeholders were originally defined by Freeman (1984, 46) as "any group or individual who can affect or is affected by the achievement of the organization's objectives" and included shareholders, employees, customers, suppliers, and society. The World Bank (2001), as part of its work on stakeholder analysis for policy work, noted a stakeholder could be of any form, size, and capacity and the stakeholder could be individuals, organizations, or unorganized groups. According to the World Bank, there are four major attributes that are important for stakeholder analysis: (1) the stakeholder's position on the issue, (2) the level of influence or power the stakeholder has, (3) the stakeholder's level of interest in the specific issues, and (4) the coalitions or groups to which the stakeholder belongs. Stakeholders are those ultimately affected, either positively (beneficiaries) or negatively. Stakeholders include both winners and losers, and those involved or excluded from the decision-making process (Overseas Development Administration 1995). The stakeholders are insiders for a situation, and an objective of an RQI is to get at their perspective or understanding of a situation. The individuals the RQI team collects information from are stakeholders, but are unlikely to be all of the stakeholders. To some extent, the individuals with whom the team interacts represent an opportunity sample of the stakeholders.

The sponsors often provide the funding for the RQI and initiate the arrangements; they are the clients to whom the RQI team is responsible. Often the sponsors are also stakeholders, since the results may have an impact on them. There are also some situations in which the local stakeholders sponsor the RQI, and the RQI team works directly for and is responsible to the stakeholders. The boundaries between these groups are permeable and individuals can belong to several groups at the same time. The reason that the distinction between the groups is important is that interaction between the groups is influenced by differences in power and self-interests. Even when the groups are acting in good faith with each other, and my assumption is that this is almost always the case, there can be problems.

Unequal power relationships both within the RQI team and between the RQI team and the local stakeholders can seriously threaten the process. RQI requires intensive and frank interaction between team members. It also requires that local people are willing to share their concerns and categories with outsiders. Unequal power relationships can have the consequence of the less powerful concluding that their best strategy is to tell the more powerful what they think the more powerful want to hear (Leurs 1997, 292).

> The less powerful may conclude it is in their best interest to tell the more powerful what they think the more powerful want to hear.

Recognition of the unequal power between the different groups involved in an RQI can help all parties identify the limits on the independence of the RQI team. The sponsors have the power to either engage or not engage the services of the team. The sponsors may explicitly identify topics they want covered or topics that are "off limits." Often the sponsors control the release of any report prepared by the team, and they may reserve the rights to review or edit the report. In some situations, the RQI team may claim authorship of the report submitted to the sponsors and the sponsors may use this report to prepare a report that is clearly their own. The ability of the sponsors to fund additional research can have a subtle, but powerful, impact on the results of the RQI. As the RQI team attempts to increase the "usefulness" of their report, the sponsors can have tremendous influence on the report, since the sponsors define usefulness.

> Unequal power can limit the independence of the RQI team.

RQI teams are generally ad hoc, with team members selected because of what they can bring to the team. It has been my experience that team members are often selected from institutions of higher education, the sponsoring organization, or organizations involved in similar activities as the sponsoring organization. Organizations may set requirements that have to be met before their employees can engage in research. These can range from requiring a review of the research design by a committee formed to ensure the protection of human subjects such as an Institutional Review Board (IRB), to a claim on the ownership and control of work resulting from the participation of their employee on a team. RQI team members from higher education will be influenced by how their organizations define legitimate research and the pressure in higher education to publish results. The quest for promotion to higher academic ranks and for tenure puts pressure on the individual to produce results that can be published in peer-reviewed journals. Peer-reviewed journals tend to place value on the role of theoretical/conceptual frameworks for the analysis of research and what research can contribute to theory. Usually these types of issues are of very little or no interest to local stakeholders or sponsoring organizations. Because of the continued emphases in higher education on individual work, as opposed to teamwork, team members from higher education may feel pressured to produce individual reports in addition to their contribution to the team effort.

> The home organization of RQI team members can influence their independence.

The influence of the market on employees of private firms is easier to recognize than the more subtle influence of major donors on not-for-profit organizations or institutions of higher education. The religious affiliation of not-for-profit organizations and institutions of higher education may also influence the types of research their employees can undertake.

Constraints on the independence of the RQI team should be recognized and may need to be discussed at the outset with the local stakeholders. If these constraints have had a significant impact on the results, this should be referenced in the final report. Factors that have significantly limited the independence of the RQI team should be added to the RAP Sheet.

Promising More Than Can Be Delivered

It is easy for researchers to promise to deliver more than is possible and for sponsors to demand more than is reasonable. In some cases, unrealistic expectations appear to develop spontaneously. These issues are intensified by the nature of RQI. Some practitioners of RQI have been uncritical of the approach and have failed to recognize its limitations. Chambers (1991, 531) cautioned about overselling rapid research methods and suggested this provides grounds for discrediting these methods.

> The RQI team should be careful not to promise more than it can deliver.

RQI has been identified as an appropriate approach for inquiry when there is a need to get at the insiders' perspectives. Specific techniques for RQI are selected from qualitative research techniques that have proven effective in getting at the emic understanding of local situations. It must, however, be recognized that the time constraints on RQI and a limited role for participant observation preclude the depth of understanding that long-term fieldwork is capable of producing. In many situations, carefully done RQI can be expected to produce more of an insider's perspective than alternatives such as questionnaire research and can be expected to produce emic understanding on a very limited number of issues. It is not unreasonable to expect RQI to help define issues from the perspective of local participants, to provide special terms and definitions as used locally, and to identify priorities based on local criteria. However, even more than traditional qualitative research, RQI can produce results that appear to be superficial. Wolcott's comments about traditional qualitative research also would appear to apply to RQI: "Tighter conceptualization, cautious labeling, and a careful paper trail still seem the best protection against the inevitable charge that certain aspects of our work are superficial" (Wolcott 2005, 121).

In some situations, the RQI team has relatively more power in defining appropriate expectations than does the sponsor, because the team has relatively more information on the approach. I strongly believe the team has a responsibility to clearly identify the limits of RQI.

Of growing concern to many practitioners are the requests from sponsors to use rapid research methods in inappropriate situations. As noted earlier at several points, RQI is usually not appropriate when estimates of numbers or percentages are needed. Also, as noted earlier, there are social/cultural situations in which team-based research will be inappropriate. There may be times when the power of the sponsor to provide the funding has to be countered with the power of the RQI team to provide information about the limits of the research approach. If information is not sufficient to change the views of the sponsors, it becomes the responsibility of the team to refuse to try to implement RQI where it would be inappropriate.

Who Benefits?

The assumption is that social science research is done to benefit the stakeholders, either directly or indirectly, through the production of knowledge that will in some way have a positive impact. RQI is even more closely tied to the assumption of providing benefits to stakeholders than is traditional research. However, RQI, like traditional research, often provides benefits not just to the local stakeholders but to the researchers, the sponsors, and the academic community.

> RQI benefits the team, the sponsors, and the academic community as well as the local community.

The RQI Team

Everyone I know of who has participated on an RQI team has described it as hard work, yet I have not heard any former participants indicate they would not like the opportunity to participate on another team. The question is what motivates individuals to take on this task. The most frequent response that I have heard mentioned is the intrinsic reward from working with others and providing assistance. Participation on an RQI team often provides a welcome break from regular activities and, in some cases, an opportunity to travel. Participation may result in monetary compensation, but I have yet to hear of participants for whom this was the motivating factor. For many individuals, both inside and outside of academia, there are rewards for doing research and completing reports or papers based on the research. Graduate students are uniquely well positioned to benefit from participation on an RQI team, since they can experiment with data collection and data analysis techniques that they may use in their own research. They also have the opportunity to develop expertise on new topics. As Wolcott wrote, "Best intentions notwithstanding, I think we must concede that the person who stands to gain the most from any research is the researcher" (Wolcott 2005, 129).

The Sponsors

While a successful intervention resulting from an RQI may provide direct benefits to the local stakeholders, it can also provide significant benefits to the sponsors. Often, rapid research methods are initiated at higher levels of the bureaucracy to better understand problems, design new activities, or monitor ongoing activities of administrative units at lower levels of the organizations. Success in the lower administrative units or the local community may translate into increased resources for the sponsors. It should be assumed that sponsors are especially sensitive to the political context of research and that these considerations may impact decisions about what research is done, who does it, and how the results are disseminated.

The Academic Community

Organizations can benefit from learning about successes and failures in other organizations. Likewise, research approaches such as RQI can be improved by knowing about factors that contribute to the success or failure of specific inquiries. This type of information needs to be collected and disseminated. Traditionally, the results of research are included in reports or published in journal articles. Reports issued by organizations often have limited distribution, and other organizations that could make use of the results have no way of knowing they exist. Material published in academic journals may reach researchers who participate on RQI teams, but are less likely to reach local organizations. The RQI team has a responsibility to ensure that both the sponsors and the stakeholders understand the importance of disseminating the results of the RQI. This may require explicit agreement on confidentiality and an appreciation of the need to change details in the report to conceal the identity of the research site. The RQI team also has a responsibility to seek nontraditional ways of disseminating the results of their work, including but not limited to the use of the Internet. Purposeful selection of one or more RQI team members from organizations similar to the one being studied is one way of sharing information between organizations.

The Local Stakeholders

Often, the explicit objective of an RQI is to better define local issues in order to provide necessary external resources or change structural or policy constraints. One of the reasons that RQI needs credibility with the world beyond the local situation is to facilitate these types of actions by outsiders. However, a focus on the power of outsiders to provide resources or change policies can minimize the role of the insiders in solving their problems. Outsider researchers can perpetuate a model of dependency. The research process can leave local people feeling that they have no voice in the identification of their problems or in suggesting solutions. With sensitivity to this issue and with the role of the insider when possible as a full member of the RQI team, the RQI process can be a collaborative process of defining issues. The RQI activity can contribute to the development of local skills and systems for addressing issues that will be active after the RQI is finished. To be truly helpful to the insiders, the RQI process must help empower the local individuals while also helping them make their case for changes that can only be made by outsiders.

Issues of Propriety, Costs, and Harm

Propriety refers to using procedures that are ethical and fair to those who are involved and are affected by the results of the inquiry. Propriety requires informed consent and attention to issues that represent costs and can be harmful to participants.

Informed Consent

> Written informed consent is necessary.

I believe that getting informed consent is essential for the fair treatment of the local stakeholders. Some researchers have argued that interviews do not require written informed consent. I believe, along with most universities and funding agencies, that written informed consent is needed for interviews. Getting informed consent from participants provides an opportunity to discuss with them the purpose of the activity and to ensure that a minimum amount of agreement exists. The definition of informed consent can be expected to be culturally specific and this topic should be discussed with local key informants early in the process. When dealing with individuals who are functionally illiterate or who do not want to sign a form, one option is to have them record their consent. The form should be read to them and they should be asked to say either "yes" or "no" to indicate they agree with the terms. This should be played back to the respondent before the interview proceeds (and can be part of the testing of the recording equipment). Schoolchildren, other minors, and individuals with guardians require the formal, informed consent of their parents or legal guardians. Informed consent is premised on the individual having the freedom to decline without fear of retribution, a condition that individuals in institutions like prisons or even schools may not feel they have. Figure 8.1 is a sample informed consent form that can be adapted for specific situations. Universities and funding agencies may have their own forms that must be used.

Value of Time

The time people spend responding to the questions of an RQI team is valuable and the team should recognize this. One way to recognize the value of someone's time is to arrange meetings when it suits the respondents. Recognizing and appreciating the value of someone's time is not the same as paying people to be interviewed. Generally, payment is to be avoided.

Betrayal

Researchers rarely set out to betray the people who are the subjects of their inquiry, yet local participants are sometimes betrayed and often feel betrayed. Miles et al. suggested that research can be considered an act of betrayal since "you make the private public and leave the locals to take the consequences" (Miles et al. 2014, 297). Betrayal is most likely to occur when the research team promises a level of confidentiality that is not delivered and when the research discloses information that should not be made public.

The process of providing informed consent to participate in any inquiry may include agreement that confidentiality will be provided. There are two problems with promises of confidentiality. First, there may be significant cultural differences in how confidentiality is defined. This is likely to be more of a problem in the types of situations where RQI is needed than in other research situations. An informed consent form can have a very different meaning depending upon whether the reader of the form is the researcher or the researched. Even when there is agreement on the meaning of a concept like confidentiality, there is very real danger to the

RQI Informed Consent Form

RQI team leader or representative:
Name: ____________________ Contact information: (phone, email) __________________
Address: ____________________
If you have questions or concerns about this activity, please feel free to contact
Name: ____________________ Contact information: (phone, email) __________________
Position: ____________________
Thank you for agreeing to participate in this Rapid Qualitative Inquiry (RQI). This form outlines the purposes of the study and provides a description of your involvement and rights.
This is a qualitative study based on the use of a small team of researchers who will conduct semistructured interviews and combine these with observations and information collected in advance. The purpose of this activity is to get your opinions, insights, and suggestions about:

____

____
You are encouraged to ask questions at any time about the activity and the methods we are using. Your suggestions and concerns are important to us. We will use the information from this activity to write a report. The report will be a public document. Please let the RQI team know if you would like access to the results of this activity.
Unless you are asked to sign a separate statement at the bottom of this form, your real name will not be used at any point in the written report. Instead, you and any other person and place you name will be given fictitious names (pseudonyms) that will be used in all verbal and written records and reports. To the best of our ability, and consistent with law, what you share with us will be kept confidential unless you agree otherwise. Audio files of interviews will be used only for this study, will not be played for any reasons other than to do this study, and will be erased or destroyed within three years following the completion of the report.
Your participation in this study is voluntary. You have the right to withdraw at any point of the study, for any reason. If you withdraw, information collected from you and records and reports based on information you have provided will not be used.
Do you agree with the terms of this agreement? ______ (please write in YES or NO)
Your name (printed) __________________________
Your signature ______________________________ Date ______________
==

Special Consent to Use My Real Name

There may be special circumstances in which the use of your real name is desirable, such as the need to add credibility to statements or indicate support for proposed actions suggested by the community. If there is a possibility that your real name will be used, you will be asked to sign below. If your real name is used, you will be given the opportunity of reviewing the draft report and suggesting changes before it is finalized.

Do you grant permission for the use of your real name? ___YES ___ NO

Your signature ______________________________ Date ______________

Figure 8.1. Sample Informed Consent Form

participant if the researcher is naive and falsely believes that it is possible to guarantee confidentiality. Accidental disclosure or forced disclosure by legal authorities can threaten confidentiality. An often overlooked source for accidental disclosure is the individual who transcribes the interview tapes. This is especially true with individuals for whom transcribing interviews is not their primary responsibility. Concern about this issue by several of the participants in the community college RAP/RQI resulted in the requirements that some of the interviews be transcribed by individuals not associated with the college and that those involved in the transcription process sign a nondisclosure form. Figure 8.2 is a sample transcriber nondisclosure form that can be adapted to local conditions.

The RQI team is cautioned to let the participants in the inquiry understand that the team cannot guarantee confidentiality, but that they are committed to making a good faith effort if the participants desire confidentiality. The RQI team should never promise more confidentiality than can be delivered.

The second situation in which betrayal may occur is when the results are reported. I fully agree with Wolcott that "no fieldworker ever has a license to tell all" (2005, 141). Wolcott suggested that an indication of the success of fieldwork is learning things that you did not want to learn and then facing decisions about what to disclose, "at what cost, for what audiences" (142).

> I think fieldworkers should always have in mind the boundaries of their inquiries. . . . You have to respect people's efforts to convey what they want you to hear, just as you hope they will talk candidly about what you want to hear. Your sense of courtesy will guide the extent to which you allow them just to go on and on. The limits set for the inquiry should guide the extent of extraneous materials to record. (142)

> Betrayal occurs when researchers incorrectly believe they have a license to tell all.

Bogus Empowerment

Anytime people are led to believe they will have the authority and power to implement changes when in fact this is not the case, they are experiencing bogus empowerment. An RQI team that lets the respondents

Rapid Qualitative Inquiry-RQI
Transcript Confidentiality Agreement

Thank you for agreeing to transcribe audio files of interviews for a Rapid Qualitative Inquiry (RQI). This form outlines the purpose of the study and provides a description of expectations about your involvement.
The purpose of this activity is to get participant's opinions, insights, and suggestions about ____

____

We have promised the participants that their real names (and the real names of anyone they refer to) will not be used at any point in the final written report, unless we have their written approval to use their real names. All participants will be given fictitious names (pseudonyms) that will be used in all verbal and written records and reports. We have also promised them that audio files of interviews will be used only for this study and will not be played for any reasons other than to do this study.
The ability of the team to accurately convey the statement of the participants depends to a large extent on your ability to accurately transcribe the tapes. If you have any questions about what is being said, please indicate it in the transcript by typing ??? (three question marks).
In agreeing to transcribe the interviews, you agree not to reveal their contents or even the names of the individuals interviewed.

Do you agree with the terms of this agreement?

PLEASE WRITE IN YES OR NO ___________

Your name (printed) ________________________________

Your signature ________________________________

Date ________________________________

Figure 8.2. Sample Transcriber Nondisclosure Form

in their inquiry believe either they will be given the power and authority to implement the results of the RQI or that the recommendations of the RQI will likely be implemented by someone else, when there is neither a structure nor plans for implementing the RQI recommendations, is being dishonest and is guilty of bogus empowerment.

Ciulla introduced and defined the concept of bogus empowerment (see Ciulla 2003, 2004). According to Ciulla the difference between authentic and bogus empowerment depends on the honesty of the relationship between the individuals involved (2004, 76). "Honesty entails a set of specific practical and moral obligations and is a necessary condition for empowerment" (76). Her observation that decisions by even those with the best of intentions are dominated by differences in power and can lead to bogus empowerment is especially relevant to RQI. Ciulla explained that when someone really empowers someone else, the empowered person is also given the responsibility that comes with that power (77). She went on to note that those who are empowering others have to keep their promises, and that the best way to do this is to make promises that can be kept. Ciulla cautioned about the need to avoid "the temptation of engaging in hyperbole about the democratic nature" of organizations and motivation for specific actions (78).

She noted that "empowerment is about giving people the confidence, competence, freedom, and resources to act on their own judgments" (2004, 59) and that bogus promises of empowerment are "cruel and stressful" (78).

Two aspects of authentic empowerment identified by Ciulla are especially relevant to RQI. Both the RQI team and those responsible for bringing in the RQI team must keep their promises. The second aspect of authentic empowerment especially relevant to RQI is the need to "overthrow some of the aspects of niceness" (Ciulla 2004, 79). Ciulla observed that "the truth is not always pleasant" and concluded that "when you really empower people, you don't just empower them to agree with you" (79).

The history of the Student Services Division at the community college had produced a culture characterized by mistrust. Experience had taught that little should be expected from efforts like the RAP/RQI. In this case, the ability of the participants to trust the RAP/RQI team depended to some extent upon their ability to trust the new dean. The new dean, however, hoped that the RAP/RQI process could help build some of the trust that was needed. The situation called for complete honesty about motives and processes involved in requesting the RAP/RQI team. The first meeting with the leadership team of the division provided the opportunity for the needed honesty. Without this honesty, the results could have been what Ciulla (2004) calls "bogus empowerment."

For the community college RAP/RQI, both the team and the dean were careful to clearly indicate that, even though all voices would be listened to, not all recommendations could or would be acted upon. Both the team and the dean also indicated that some issues would be identified that were beyond the influence of the dean, but that where appropriate these issues would be included in the report.

> RQI can be a facilitator of bogus empowerment by encouraging people to falsely believe that their input will be acted upon.

The agreement between the RAP/RQI team and the dean was that the RAP/RQI team would have editorial control over the report. This information was communicated to the employees in the division. The report included controversial issues and issues for which there were no easy or even likely answers. It was also communicated to the members of the division that the RAP/RQI team planned to use the data from the study for scholarly activities as well as for whatever use the college wished to make of them. Finally, agreement on confidentiality was made, redefined, and reaffirmed on numerous occasions.

Institutional Review Boards

An Institutional Review Board (IRB) is a committee of five or more individuals with the responsibility of reviewing research proposals and monitoring ongoing studies to ensure the protection of human research subjects. Most research projects, including RQIs, must receive approval from their institution's IRB before they start. Any study that involves living human subjects needs IRB approval and while IRB rules exempt some types of research, almost all studies need to be able to show as part of their IRB application that attention has been given to protecting human subjects.

IRB reviews can take anywhere from several weeks to months and the time for the review should be included in the planning process. IRBs may not be familiar with qualitative research such as RQIs and this can increase the time needed for approval. The most critical elements of the IRB application may be a careful description of the activity and the informed consent form that participants in the activity will be asked to sign. Jo Anne Schneider has written extensively about Institutional Review Boards and qualitative research (see Schneider 2013). Schneider noted that informed consent should (a) tell participants that their participation is voluntary, (b) ensure them that their privacy will be maintained, (c) inform them of the plans for carrying out the research, (d) explain why the project is being done, and (e) describe how the information will be used. Informed consent letters must include information on whom to contact for additional information and whom to contact if the participant has concerns about the research being conducted. The sample Informed Consent Form in Figure 8.1 is based on a form prepared by Gonzaga University and covers the requirements identified by Schneider.

Most IRB applications for RQIs will indicate that there is minimal risk to participants and should not indicate that there is no risk. Schneider defined minimal risk as a situation in which "the probability and magnitude of harm or discomfort anticipated in the research are not greater than those ordinarily encountered in daily life or during the performance of routine physical or psychological examinations or tests."

Oakes (2002, 454) recommended that one approach to dealing with review boards not familiar with qualitative research is for the researcher to "educate himself or herself about IRBs and educate IRBs about research." Oakes further recommended researchers need to educate themselves about specific IRB regulations and issues and to then volunteer to serve on the IRB.

Different institutions will have different requirements for research carried out as classroom activities or during training programs. Some institutions may require an annual IRB application for all anticipated teaching-related research activities and then copies of letters of informed consent for individual activities. It is important that local requirements are fully understood.

General Ethical Concerns

This chapter has highlighted several issues that are particularly relevant to RQI, whether it is being used by practitioners or by students in a class. A consideration of all of the general ethical issues involved in doing RQI is beyond the scope of this book. I believe, however, that the "Statement of Professional and Ethical Responsibilities" of the Society for Applied Anthropology provides a useful summary of issues that should be considered. This statement is reproduced in figure 8.3. Practitioners of RQI are reminded of the statement of Mirvis and Seashore (1982 as cited in Miles and Huberman 1994, 288): "Naiveté [about ethics] itself is unethical."

Society for Applied Anthropology

This statement is a guide to professional behavior for the members of the Society for Applied Anthropology. As members or fellows of the society, we shall act in ways consistent with the responsibilities stated below irrespective of the specific circumstances of our employment.

1. To the peoples we study we owe disclosure of our research goals, methods, and sponsorship. The participation of people in our research activities shall only be on a voluntary basis. We shall provide a means through our research activities and in subsequent publications to maintain the confidentiality of those we study. The people we study must be made aware of the likely limits of confidentiality and must not be promised a greater degree of confidentiality than can be realistically expected under current legal circumstances in our respective nations. We shall, within the limits of our knowledge, disclose any significant risks to those we study that may result from our activities.

2. To the communities ultimately affected by our activities we owe respect for their dignity, integrity, and worth. We recognize that human survival is contingent upon the continued existence of a diversity of human communities, and guide our professional activities accordingly. We will avoid taking or recommending action on behalf of a sponsor which is harmful to the interests of the community.

3. To our social colleagues we have the responsibility to not engage in actions that impede their reasonable professional activities. Among other things, this means that, while respecting the needs, responsibilities, and legitimate proprietary interests of our sponsors we should not impede the flow of information about research outcomes and professional practice techniques. We shall accurately report the contributions of colleagues to our work. We shall not condone falsification or distortion by others. We should not prejudice communities or agencies against a colleague for reasons of personal gain.

4. To our students, interns, or trainees, we owe nondiscriminatory access to our training services. We shall provide training which is informed, accurate, and relevant to the needs of the larger society. We recognize the need for continuing education so as to maintain our skill and knowledge at a high level. Our training should inform students as to their ethical responsibilities. Student contributions to our professional activities, including both research and publication, should be adequately recognized.

5. To our employers and other sponsors we owe accurate reporting of our qualifications and competent, efficient, and timely performance of the work we undertake for them. We shall establish a clear understanding with each employer or other sponsor as to the nature of our professional responsibilities. We shall report our research and other activities accurately. We have the obligation to attempt to prevent distortion or suppression of research results or policy recommendations by concerned agencies.

6. To society as a whole we owe the benefit of our special knowledge and skills in interpreting sociocultural systems. We should communicate our understanding of human life to the society at large.

Figure 8.3. SfAA Statement of Professional and Ethical Responsibilities

In the next chapter I will examine Rapid Qualitative Inquiry's history and relationship to selected other research methodologies. The focus will be on what RQI shares with some of the other approaches, as opposed to differentiating RQI from them. RQI is flexible enough that it can be made more relevant to specific circumstances by borrowing from other approaches.

Additional Reading

Chapter 3, "Ethical Issues in Analysis," in Miles et al. (2014), is a comprehensive review of issues. Chapter 6, "Fieldwork: The Darker Arts," in Wolcott (2005), raises important issues that I have not seen covered elsewhere. Ciulla (2004) provides a label for something many of us have experienced and all of us wish to avoid doing to others: bogus empowerment. Schneider (2013) is a comprehensive guide to qualitative research and IRBs. If ordered online, the cost includes a sixty-minute webinar. Some of the most critical elements of her guide can be found by doing a Google search.

Ciulla, Joanne B. 2004. Leadership and the problem of bogus empowerment. In *Ethics: The heart of leadership*, edited by J. B. Ciulla, 63–86. Westport, CT: Praegar.

Miles, Matthew B., A. Michael Huberman, and Johnny Saldaña. 2014. *Qualitative data analysis: An expanded sourcebook.* 3rd ed. Thousand Oaks, CA: Sage.

Miller, Tina, Maxine Birch, Melanie Mauthner, and Julie Jessop. 2012. *Ethics in qualitative research.* 2nd ed. Los Angeles, CA: Sage.

Punch, Maurice. 1986. *The politics and ethics of fieldwork.* Beverly Hills, CA: Sage.

Schneider, Jo Anne. 2013. *Qualitative research and IRB: A comprehensive guide for IRB forms, informed consent, writing IRB applications and more.* Bonita Springs, FL: Principal Investigators Association.

Wolcott, Harry. F. 2005. *The art of fieldwork.* 2nd ed. Walnut Creek, CA: AltaMira.

CHAPTER NINE
RAPID RESEARCH AND THE RQI FAMILY TREE

Main Points

1. Most rapid research methods can be described as "first-cut assessments of . . . poorly known areas" (Conservation International 1991, 1), and as "organized common sense, freed from the chains of inappropriate professionalism" (Chambers 1980, 15).
2. The common thread in the evolution of rapid researchmethods is the conclusion that the world was moving too quickly for the normal, disciplinary approaches.
3. Despite their different origins, many rapid research methods have in common the fact that they were introduced to address the need for cost-effective and timely results in rapidly changing situations.
4. RAP and RQI have their roots in farming systems research but other forms of rapid qualitative research have other roots.
5. RQI is a direct descendant of Rapid Appraisal, Rapid Assessment, Rapid Rural Appraisal, and Rapid Assessment Process (RAP).
6. RQI represents a middle ground that is far more participatory than early versions of Rapid Appraisal, but not necessarily as participatory as Participatory Rural Appraisal (PRA).
7. Even though RQI is a form of RAP, not everything referred to as "Rapid Appraisal" or "Rapid Assessment" meets the methodological requirements of RQI.

CHAPTER NINE

Rapid Qualitative Research

The objective for most of the approaches that are identified as rapid research methods could be summarized as "first-cut assessments of . . . poorly known areas" (Conservation International 1991, 1). The element of compromise is explicit in statements that describe them as "a middle zone between . . . fairly quick and fairly clean" (Chambers 1991, 521). Rapid Appraisal was described as "organized common sense, freed from the chains of inappropriate professionalism" (Chambers 1980, 15) and "a form of appropriate technology: cheap, practical and fast" (Bradfield 1981 cited in Rhoades 1982, 5). These last two comments are especially relevant to RQI.

> The objective for most rapid research is "first-cut assessment of . . . poorly known areas."

Proliferation of Terms for Rapid Qualitative Research

The proliferation of terms used to describe rapid research methods and the use of similar terms for different approaches complicates any discussion of rapid research methods. The goal of this chapter is to simplify the discussion by providing some context and history of some of the different approaches. I recognize that the introduction of yet another label, Rapid Qualitative Inquiry, has the potential to only further complicate this discussion, but I am convinced the potential benefits are greater than the potential costs. Users will continue to call the methods they use by whatever name they believe is most appropriate, regardless of what others call it.

Early History of Rapid Qualitative Research

In 1941, Robert Redfield and Sol Tax conducted a three-day field survey and used the term *Rapid Guided Survey* (Wolcott 2005, 103). Despite having a clear idea of the information they sought, since they were in the seventh year of their fieldwork, they attributed their success with the survey at least in part to sheer luck.

In the early 1970s Spradley and McCurdy (1972) introduced an approach for teaching undergraduate students about ethnographic research based on doing "cultural descriptions" of "cultural scenes." Their book covered ways of discovering cultural categories, organizing them into larger domains, and identifying the elements or attributes that give them meanings. They estimated that the total amount of time needed for the entire process was about the same time as it should take an undergraduate student to do a library research paper. They argued that "an adequate study can be done in a limited amount of time by restricting the scope of your investigation" (5). They also pointed out that even a researcher spending two years in the field can examine only a few topics in depth and also has to restrict his or her investigation. They suggested that the twelve ethnographic reports done by their students and included in their book "demonstrate the way cultural scenes in our own society can be studied" (ix). These themes were expanded in their second edition (McCurdy, Spradley, and Shandy 2005) and focused on the concept of microcultures, a concept that can be applied to most rapid research methods. As noted earlier Margaret Mead is described as being neither bashful nor apologetic about the short duration of some of her fieldwork (Wolcott 1995, 78).

Multiple Origins for Rapid Research

Fitch, Rhodes, and Stimson (2000) noted that rapid research methods have appeared in agriculture, community development, rural development, marketing, and health care. Despite their different origins, they were all introduced to provide cost-effective and timely results in rapidly changing situations (64). According to Fitch et al., one area where the use of rapid research methods has been extensive and has been documented is the public health field of drug and alcohol and that beginning as early as the 1970s rapid research methods have been used to inform drug policy. Manderson and Aaby (1992 as cited by Fitch et al. 2000, 63) referred to an "epidemic" in the use of Rapid Assessment methods. Fitch et al. noted that like an actual epidemic its "progression has been shaped through its exposure to different cultural, institutional, political, policy, and public health environment" (63). Fitch et al. indicated that historically Rapid Assessment was typified by diversity and uncertainty and in its broadest sense denoted approaches that aimed to be cost effective, timely, and inductively informed by a range of methods. In the substance abuse field

Rapid Appraisal was a response to large-scale, quantitative, and centralized research studies that took years to produce results. In addition to the time, early large-scale studies relied on predetermined definitions of risk that turned out to be problematic. First attempts at more rapid approaches in the substance abuse field were poorly documented. The next stage was the development of "Rapid Assessment Methodology" (RAM) guidelines followed by the development of Rapid Assessment and Response (RAR) guidelines. RAR guidelines were used extensively in studies worldwide. Fitch et al. (74) noted the influence of Rapid Rural Appraisal and Rapid Assessment Procedures (RAP) on the development of rapid approaches in the substance abuse field. They noted the need for building alliances and sharing results of rapid approaches in order to improve their clarity and "scientific status." They also noted a need for wider publication and dissemination of results, and for moving beyond the association of rapid approaches with developing-country use.

Farming Systems Research and Rural Studies

Farming systems research, as experimented with in developing countries in the late 1970s (see Shaner, Philipp, and Schmehl 1982), provided my introduction to rapid research. Farming systems research was based on a consideration of people along with their plants and livestock. There was an increased need to know about the conditions farmers faced when they carried out agriculture, and neither research tourism nor questionnaire survey research could provide solid and timely results. Photocopies of papers about new approaches for research made their way to projects around the world, including Sudan, where I was assigned at the time. I was especially impressed with a paper by Peter Hildebrand (1979) describing a research approach based on teamwork called "**Sondeo**." I then received copies of several other papers on rapid, team-based research from the Workshop on Rapid Rural Appraisal, held October 26–27, 1979, at the Institute of Development Studies, University of Sussex, including papers by Robert Chambers. The title of this workshop used the phrase "Rapid Rural Appraisal." This phrase and variants of it became associated with rapid, team-based research. Despite differences in details, people generally used the phrase Rapid Rural Appraisal for activities based on small multidisciplinary teams using semistructured interviews to collect information and completing the entire process in weeks or a few months.

Rapid Rural Appraisal (RRA) and Rapid Assessment

RQI is a direct descendant of Rapid Rural Appraisal as well as Rapid Appraisal and Rapid Assessment. One of the strengths of these approaches has always been their flexibility. These approaches have continued to evolve and the terms associated with them have been used in different ways, to describe a wide range of activities. As suggested by the title of this book, I believe RQI is Rapid Assessment with the critical distinction of always being team based. I know that not everything that is labeled as Rapid Rural Appraisal, Rapid Appraisal, or Rapid Assessment would meet the methodological requirements of RQI.

> Not everything called RAP meets the methodological requirements of RQI.

Chambers (2008, 70) described the development of Rapid Rural Appraisal as a response to the "tyranny of methodological rigidities" of the 1970s. According to Chambers social anthropologists "believed that only their approach could yield in-depth understandings. Economists and statisticians believed that only questionnaires could generate the numbers needed in rigorous research. The researchers and practitioners who improvised outside these traditions were a heretical fringe and the methods disparagingly dismissed as 'quick-and-dirty.'" The approach and methods that were improvised outside the traditional methods came to be known as Rapid Rural Appraisal (RRA). Rapid Rural Appraisal addressed the need for information by decision makers that was relevant, timely, accurate, and usable. Rapid Rural Appraisal was described by Chambers (72) as a response to "development tourism," the brief rural visit by the urban-based professional. Problems with development tourism included misleading replies, failure to listen, reinforced prejudice, a focus on the visible as opposed to the invisible, and seeing snapshots and not trends. Alternatives to development tourism were identified by Chambers as "long and dirty" approaches and included "ritual immersion in alien cultures," "huge questionnaire surveys," and a collection of maps of soil, vegetation, land use, and rainfall that "sits on the shelf," "interesting" but of "no use to planners."

Rapid Rural Appraisal was described as a middle zone, "a zone of greater cost-effectiveness" (73). According to Chambers the two principles for Rapid Rural Appraisal were (a) optimal ignorance, described as knowing "what is not worth knowing" and (b) proportionate accuracy described as avoiding "accuracy which is unnecessary" but rather looking at orders of magnitude and directions of change (74).

Two principles: Optional ignorance or knowing what is not worth knowing and avoiding accuracy which is not necessary.

General principles for RRA suggested by Chambers include: taking time, talking about biases, being unimportant and avoiding the limousine, listening and learning, and using multiple approaches (74). Chambers offered ten "disparate techniques" as illustrations of the "range of possibilities": (a) use of existing information, (b) learning indigenous technical knowledge, (c) using key indicators, (d) adaptation of Hildebrand's "Sondeo" approach, (e) use of local researchers, (f) direct observation, (g) use of key informants, (h) group interviews, (i) use of the guided interview, and (j) use of aerial inspections and surveys.

Chambers concluded his discussion of Rapid Rural Appraisal by noting it "is no panacea" and is subject to "superficiality and error," with time being "the most critical factor" (79).

Participatory Rural Appraisal and Related Participatory Approaches

Chambers (2008, 86) noted that beginning in the mid-1970s there was an "accelerating evolution" of participatory methodologies in development practice that led to the development of **Participatory Rural Appraisal (PRA)** and Participatory Learning and Action (PLA). Chambers suggested that many practitioners had become "eclectic methodological pluralists" (86) and that the future of rapid participatory approaches lies not in branding and boundaries but in eclectic pluralism (85).

The future of rapid research requires eclectic pluralism.

Absalom et al. (as cited by Chambers 2008) identified characteristics of PRA methods. They were usually "performed" by small groups and were "visible." Methods often included maps and diagrams made by local people and methods could include timelines, trend diagrams, and proportional piling of cards or objectives. A second characteristic of PRA was attitudes and behavior changes. The third aspect of PRA was sharing, focusing on knowledge, but also including food, ideas, insights, and relationships, and "sharing without boundaries."

Chambers (2008) noted there was "much intermingling and innovation" in the origins of PRA but that agro ecosystem analysis and Rapid Rural Appraisal were central. Agro systems analysis was credited with contributing mapping and Rapid Rural Appraisal was credited with contributing semistructured interviews. According to Chambers the major "breakthrough" in the development of Participatory Rural Appraisal was the rediscovery that local people had the ability for "action, experimentation, research, and mentoring and evaluation" (88). The most important contribution of Participatory Rural Appraisal has been enabling the marginalized and weak to do their own appraisals and analysis and to gain voice and to take their own action (102). Chambers also noted that the label Participatory Rural Appraisal had become a problem and was being used to describe procedures that lacked PRA principles and practices (101).

Chambers identified Rapid Assessment Procedures (see Scrimshaw and Gleason 1992) and the Rapid Assessment Process (Beebe 2001, the first edition of this book) as contributing to the continued development of Rapid Rural Appraisal and by implication contributing to the development of Participatory Rural Appraisal. I appreciate Chambers's description of the first edition of *Rapid Assessment Process* as "substantial, authoritative, and useful" (Chambers 2008, 81).

RQI and Participatory Research

Some of the advocates of **Participatory Rural Appraisal** (PRA) would probably say that RQI is not "participatory." They might argue that RQI is closer to traditional Rapid Appraisal, with its focus on data collecting, analysis done mainly by outsiders, and dependence on observations and semistructured interviews. They would contrast this with PRA, with its focus on enabling local people to share, embrace, and analyze their knowledge of life and to plan and act based on mapping, diagramming, and comparison. For PRA the assumption is that local people will do almost all of the investigation and analysis and will then share with the outsiders (Chambers and Blackburn 1996). Advocates of PRA refer to "the use of local graphic representations created by the community that legitimize local knowledge and promote empowerment" (CASL 1999).

> RQI is designed to be more participatory than early Rapid Appraisal but not as participatory as PRA.

I would argue that RQI represents a middle ground that is far more participatory than early versions of Rapid Appraisal but not as participatory as Participatory Rural Appraisal. There can be local representation on the RQI team. RQI is based on data collection and analysis by the team, including whenever possible an insider on the team. RQI makes extensive use of drawing and diagramming by the local community for both information collection and analysis. RQI identifies these as optional techniques to supplement semistructured interviewing and direct observation. RQI does not necessarily focus as much on empowering as does PRA. However RQI can and should empower local people. The differences in approaches are matters of degrees. The real difference is the audience. RQI is based on the assumption that local problems often reflect structural conditions over which local people have little or no control. The changes that are needed may require externally supported interventions, policy changes, or resources. Support for these types of changes requires research methodologies with credibility with the outsiders. RQI is designed to have this type of credibility. There are times when an RQI is more appropriate, times when a more participatory PRA or related approach is more appropriate, times when both are needed, and still other times when the most appropriate methodology is a combination of the two.

RQI assumes local problems often reflect structural conditions over which local people have little or no control.

RQI and Ethnography

Approaches to rapid qualitative research share many of the characteristics of an approach to **qualitative research** developed by anthropologists and referred to as ethnography.

In 1980 Spradley suggested labels for various levels of ethnography. These labels included "macro-ethnography" for the study of a complex society and "micro-ethnography" for the study of a single social situation (30).

Handwerker (2001) introduced the term *Quick Ethnography* to describe an approach for doing "high-quality" ethnographic research in between thirty and ninety days. His approach is based on starting with a clear vision of the goals of the research and paying close attention to how to achieve them. This focus is combined with techniques for making the best possible use of time in the field. Handwerker noted that this approach assumes the principal researcher is well trained in qualitative research and has sophisticated computer skills, including the use of quantitative and qualitative data analysis computer programs.

Numerous rapid research methods are identified as ethnographic or ethnography.

Rapid Ethnography as an Alternative Label for RQI

I considered using the phrase *Rapid Ethnography* instead of *Rapid Qualitative Inquiry* because ethnography (and case study), describes the methodology. Ethnography is based on data collected from documents, participant observation, and semistructured interviewing. Ethnography focuses on observable and learned patterns of behavior by a social group or individuals within the group. Ethnography is capable of generating a cultural portrait based on observations and listening to the voices of informants. All of the statements above apply to RQI as well as to ethnography (the relationship of ethnography to RQI was explored in the last half of chapter 2).

However, for traditional ethnographers like Wolcott (2005, 103) the problem was that "terms like *ethnography* or *fieldwork* join uneasily with a qualifier like *rapid*." Ethnography is usually defined as based on prolonged fieldwork. Keesing and Strathern (1998) made the claim that for most anthropologists fieldwork is central to ethnography and that "successful fieldwork is seldom possible in a period much shorter than a year" (7). Wolcott (2005) explicitly equated ethnography with extended fieldwork and stated, "Fieldwork takes time" (69). For Wolcott the ideal of two years or longer of fieldwork had already been shortened to twelve months but twelve months as a standard was "well entrenched" (69).

In 1995 Bernard suggested that it might be possible to achieve competency in another culture in as few as three months (151). Wolcott was skeptical and responded by saying, "I hope Bernard has not inadvertently foreshortened the acceptable period for fieldwork for those who will carefully misread his statement or reassure themselves that the three months he says is adequate to establish oneself in the field is all the time one needs to devote to a study" (1995, 110).

Wolcott (2005, 39) suggested that when one is in doubt as to whether something is "genuine ethnography," one should select a more cautious term. Given near consensus that ethnography requires prolonged fieldwork, I chose the term *Rapid Assessment Process* for the first edition and *Rapid Qualitative Inquiry* for the second edition instead of *Rapid Ethnography*.

> RQI is based on ethnography and case study, but is not an approach for doing either.

Concerns by Wolcott and others have not prevented numerous authors from using the word *ethnographic* or *ethnography* to describe the specific approach for rapid research they have used. Labels used and authors include *applied ethnography* (Savage 2006); *ethnographic field approach* (Finnie et al. 2010); *ethnographic praxis* (Johnsnen and Helmersen 2009); *occupational ethnography* (Cullen, Matthews, and Teske. 2008);

quick ethnography (Handwerker 2001); *rapid ethnographic assessment (REA)* (Bentley et al. 1988; Taplin et al. 2002; Guerrero et al. 1999; Kresno et al. 1994; Carley et al. 2012; Schultz, Van Arsdale, and Knop 2009); *rapid ethnographic assessment procedures (REAP)* (Low, Taplin, and Lamb 2005); *rapid ethnographic inquiry* (Westphal and Hirsch 2010); *rapid ethnographic methodology* (Mignone et al. 2009); *rapid ethnographic needs assessment* (Sandhu, Altankhuyag, and Amarsaikhan 2007); and *rapid ethnography* (Millen 2000).

Other Approaches to Rapid Research

McNall and Foster-Fishman (2007) provided a comprehensive review of selected rapid research methods that had been developed between the mid-1970s and 2007 when they published their article. In addition to Rapid Assessment (RAP; Beebe 2001; Trotter and Singer 2005), Rapid Rural Appraisal (RRA; Chambers 1994; Rifkin 1996), and Participatory Rural Appraisal (PRA; Chambers 1994), they identified Real-Time Evaluation (RTE; Bartsch and Belgacem 2004; Jamal and Crisp 2002; Sandison 2003), Rapid-Feedback Evaluation (RFE; McNall et al. 2004; Sonnichsen 2000), Rapid Ethnographic Assessment (REA; Bentley et al. 1988; Guerrero et al. 1999; Kresno et al. 1994), and Rapid Evaluation Methods (REM; Anker et al. 1993). McNall and Foster-Fishman (2007) noted there are distinctions in the origins, methods, and contexts of practice for these approaches, but what they share in common is a similar set of techniques for putting "trustworthy, actionable information in the hands of decision makers at critical moments." For McNall and Foster-Fishman a central issue in the use of rapid research methods is "achieving a balance between speed and trustworthiness." Another rapid research method not identified by McNall and Foster-Fishman is Participatory Poverty Assessment, also called Participatory Policy Research (Robb 2002). There are many more approaches to rapid research with different names.

> There are many labels for rapid research methods.

CHAPTER NINE

Rapid Research Methods in Different Fields

Table 9.1 illustrates the wide range of topics that have been investigated using rapid research methods. Table 9.1 includes several of the thirteen studies reviewed by McNall and Foster-Fishman (2007). While most of these examples report on specific research that used

Table 9.1. Selected Examples of Rapid Research in Different Fields since 2000

	General
Handwerker 2006	Evolution of ethnographic research methods
McNall and Foster-Fishman 2007	Rapid evaluation, assessment, and appraisal
Nunns 2009	Comparison of rapid evaluation and assessment
	Agriculture and Natural Resources
Driscoll et al. 2012	Fish consumption in culturally distinct communities
Gilden 2005	Social science in the Pacific fishery management
Lam and de Mitcheson 2011	Sharks of Southeast Asia
Pomeroy and Stevens 2008	A commercial fishing community in California
Toness 2001	Appraisal approaches for agricultural extension
	Business
Gifford et al. 2010	Corporate social responsibility and gold mining
Krueger 2006	The impact of the Internet on news and music business models
	Education, Youth, and Children
Aylward et al. 2010	Targeting Australian parents of young children with attachment issues
Baldé 2004	Retention of Fulani Muslim girls in Guinea
Brown, L. 2008	Mobile learning to teach reading to ninth-grade students
Brown, M., et al. 2008	Exploring HIV/AIDS knowledge and behaviors of university students in Botswana
Mpondi 2004	Education and national identity formation in Zimbabwe
Thieme 2010	Youth, trash, and work in an African city
Walsh et al. 2005	Adult learners experience in a teacher certification program
	Health
Agar 2004	Approaches to qualitative health research
Angus and O'Brien-Pallas 2003	Evidence-based nursing practice
Ash et al. 2012	Computerized clinical decision support
Auerswald et al. 2004	Sampling hard-to-reach youth on sexually transmitted diseases
Bedford et al. 2012	Home delivery in rural Ethiopia

	Health
Daack-Hirsch and Gamboa 2011	Use of prenatal micronutrients supplements by working-class Filipino women
Finnie et al. 2010	Rapid assessment of tuberculosis care-seeking in South Africa
Inciardi et al. 2009	Prescription opioid abuse in an urban community
Morin et al. 2004	HIV-infected patients in a clinical care settings
Neuwirth et al. 2007	Patient care using a panel approach
Solomon et al. 2007	Rapid assessment of existing HIV prevention programs
Trotter and Singer 2005	Community interventions and AIDS
	Military and Security
Caligiuri et al. 2011	Developing cross-cultural competency in the military
Friedemann-Sanchez et al. 2008	Perspectives on rehabilitation of patients with polytrauma
Last 2005	Rapid assessment and security sector reform
Sayer et al. 2009	Provider perspective on treating veterans with mild traumatic brain injury
Schultz et al. 2009	Rapid ethnographic assessment in the military
Walter et al. 2010	Evaluation of total force fitness programs in the military
	Miscellaneous
Bull and Farsides 2012	Rapid assessments for consent processes for research
Pennesi 2007	Improving the credibility of weather forecasts
	Policy
Bonsa 2003	The private press and democracy in Ethiopia
Caracelli 2006	Enhancing the policy process for schools
McDonald 2009	Environmental justice in an urban area
Westphal and Hirsch 2010	Engaging Chicago residents in climate change action

rapid methods, a few are general articles about rapid research methods that make reference to specific studies. All of the examples are from applied situations in which rapid research was used to respond to the need for timely and cost-effective results. Not all of the examples meet the minimum requirements of RQI as outlined in this book. The list is not meant to be a bibliography of rapid research methods. A general, but somewhat dated, bibliography is the *Annotated Bibliography on Gender, Rapid Rural Appraisal and Participatory Rural Appraisal* (BRIDGES 1994). As of 2014, I, with the help of my graduate assistants, had identified more than 165 studies that were based on all or most of the characteristics of RQI.

Rapid research methods are relevant to many fields.

The examples are organized by broad fields and by year of publication. In some cases it was difficult to decide on the most appropriate field.

Additional Reading

Excellent material on participatory research is available from the Institute of Development Studies at their website http://www.ids.ac.uk/.

Blackburn, James, and Jeremy Holland (Eds.). 1998. *Who changes? Institutionalizing participation in development*. Bourton-on-Dunsmore, UK: Practical Action.

Robb, C. M. 2002. *Can the poor influence policy? Participatory poverty assessments in the developing world.* 2nd ed. Washington, DC: World Bank and the International Monetary Fund.

CHAPTER TEN
KEY POINTS RELATING TO RIGOR AND THE FUTURE OF RQI

Main Points

1. Choice of team members in terms of experience, background, cultural differences, gender, attitude, and power is critical for rigorous RQI.
2. Attention needs to be given to collecting and sharing of materials before the RQI begins.
3. Whenever possible the RQI team should include one or more insiders.
4. Sufficient time must be provided for team orientation with a special focus on team interviewing.
5. Logistics must be in place and must function if RQI is to be rapid. Having a person responsible for logistics may be more important than having an additional academic discipline.
6. Schedules need to be designed with sufficient flexibility to take advantage of unanticipated opportunities. The number of interviews may need to be limited.
7. Audio recorders should be expected to fail and even if they do not, there are advantages to having handwritten notes on observations and interviews.
8. Transcripts of interviews and other notes should be prepared within twenty-four hours of the observations.
9. A second interview with the same person may be more valuable than an interview with a new person.

10. The RQI team should spend as much time at the site as possible including eating and sleeping there when appropriate.
11. Data displays can facilitate the analysis process.
12. Presentations of preliminary findings during the RQI can be used to check results with participants.
13. The entire RQI team should be involved in the preparation of the report.
14. Schedules for RQI should include time off to promote good behavior.
15. Explicit attention is needed to identify lessons about the RQI process.
16. RQI practitioners are urged to share their results in academic journals and online.
17. The future of RQI depends upon not confusing rapid with rushed.
18. The future of RQI is tied to the future of qualitative research.
19. The future of qualitative research is likely to be influenced by decolonizing research, increasing flexibility, increasing attention to evidence, increasing attention to variability as opposed to averages, increasing attention to context, and expanding areas where qualitative research is used.

Rigor

Before considering the future of RQI, this chapter identifies and expands on key points that I believe are critical for the rigorous implementation of RQI. To some extent this chapter is a summary of material already presented but organized around specific suggestions to improve the process. Obviously, not all of the suggestions apply to every RQI.

Attention to decisions about implementation is critical for successful RQI since RQI is not based on a list of required techniques. As emphasized throughout this book, it is the intensive teamwork and the focus on the insider's perspective, triangulation of data from multiple data sources, and iterative analysis and additional data collection that define RQI. Techniques that are suggested are designed to be flexible. Even though these techniques have proven effective under a range of conditions, they may not be appropriate for a specific RQI, may need to be modified, or may need to be replaced with other techniques. What follows is designed to make the flexibility inherent in RQI easier to manage.

While specific techniques have proven effective for RQI, these are *not* the only ones that can be used.

Improving Rigor before Beginning the RQI

Choosing RQI Team Members

Factors that influence how a specific RQI will be implemented include, first and foremost, the team. Not only the technical expertise of the team members must be considered, but also their previous experience with research and, especially, with qualitative research. Prior experience working as team members and attitudes toward collaborative work may be as important as previous experience doing qualitative research.

The flexibility in the choice of research techniques means that a technique can be chosen because a team member has experience using it. If a team member is a skilled photographer, the RQI might make extensive use of photos. If a team member is skilled in the use of spreadsheets and financial systems, she could be expected to open lines of inquiry in these areas. Artists/graphic designers could be called upon to help respondents with the creation of rich pictures and to help the team with data displays as part of the data analysis process. There are yet to be explored roles for musicians and dramatists on RQI teams. Given that the objective of semistructured interviewing is to get the respondent to "tell a story," there may be roles on the team for storytellers to both encourage and help make sense out of these stories. The use of metaphors for data analysis may make the individual with a literary background one of the most important members of the team.

Sensitivity to cultural differences is essential.

Sensitivity to cultural differences is absolutely necessary on the RQI team. The need for sensitivity should be recognized by the team and nurtured. It should not be assumed that sensitivity to cultural differences is associated with a specific discipline or ethnic group. Someone on the team must always be able and willing to question the cultural implications of decisions. The types of situations where cultural sensitivity might be especially relevant include the identification of topics that can be discussed in public, the use of drawings, the role of elders, and the use of focus group interviews as opposed to individual, semistructured interviews.

The prior experience of team members should be carefully considered. RQI teams that include members with more experience in qualitative research, teamwork, and RQI may need less-explicit guidance on the selection of techniques (Grandstaff and Grandstaff 1987, 87). As discussed in chapter 5, teams with less experience may need a different type of leadership than teams with more experience.

Attitude

Attitude is everything for the RQI team. The ability to embrace ambiguity and collaborative approaches may be more important than the technical expertise and experience of team members. Attitudes should be carefully considered in team selection. Identifying and encouraging positive attitudes should be part of the orientation and a priority for the team leader throughout the process. The three attitudes that may be most critical for a successful RQI are recognizing (1) the RQI team does not know enough in advance to articulate the questions to be asked, (2) the RQI team does not know enough to provide the answers, but (3) the RQI team knows enough to want to empower others to solve their own problems. Chambers (1996) cautioned against assuming the researchers know what to ask and noted that "the beginning of wisdom is to realize how often we do not know what we do not know." As noted several times before, the complexity of most situations and the inability to know the categories and terms used by local people make getting people to tell stories more important than getting them to answer questions. If members of the RQI team do not know enough to be able to ask the right questions, they certainly do not know enough to impose their ideas. Yet Chambers (1996) noted that, often, teams impose their ideas, solutions, categories, and values without realizing they are doing it and that this makes it difficult for them to learn from the others. The third critical attitude is one that sustains and embraces sharing and lateral learning and is based on the assumption that finding solutions often depends upon empowering others to own and solve their own problems. For some "experts" who view themselves as problem solvers, this requires a fundamental change in attitude.

> RQI team members need to recognize:
>
> 1. they don't know enough in advance to be able to articulate questions,
> 2. they don't know enough to provide the answers, but
> 3. they do know enough to want to empower others to solve their own problems.

Chambers (2007) referred to the empowerment of the weak, women, and the poor. If their attitudes are neglected, the local community ceases being the partner of the RQI team and there is a danger that the process will become mechanical and reduced to a standardized set of techniques applied in a predetermined sequence designed to extract information (Leurs 1997, 291). At the same time, it should be recognized that some issues are beyond the control of local people. One of the roles of RQI is to provide research results that will help the local people facilitate structural changes and get resources from the outside. (See ch. 8, "Bogus Empowerment," and ch. 5, "Team Leadership.")

> RQI can provide the research results local people need to get structural changes and resources from the outside.

Materials Collected in Advance

Assignments concerning the collection of documents in advance of the RQI should be made early and arrangements should include the distribution of relevant materials to other team members. Someone should be responsible for both collecting and cataloging this material. If any of this material could be used in the final report, bibliographic information will be needed. Technology reviewed in chapter 7 can make both the collection and distribution of material easier. Information collected in advance can have a major impact on methodology, even to the extent of showing that something else is needed instead of, or in addition to, an RQI. The information collected in advance will affect the initial guidelines used for semistructured interviews. When specific information is not available prior to the study, extra time and special techniques may be required to gather it (see ch. 3, "Team Collection of Information in Advance of the RQI").

Insiders as Local Team Members

Not every RQI team will have insiders, but whenever possible insiders should be included as full members of the team. RQI team members who are local insiders should be identified as early in the process as possible. Basic information on RQI should be provided to them, and they should be included in the identification and collection of the materials to be collected in advance. They must be present for orientations. Participation of local team members after they have been identified may depend upon whether they can be released from their normal activities, and ensuring this release should be a priority for the team leader (see ch. 3, "Insiders/Outsiders," and ch. 5, "Outsiders and Insiders In Between").

> Full participation of the insider on the team will require their release from their regular duties.

Orientation

Orientation on RQI methodology is critical, and sufficient time should be scheduled for it. If team members, including the insider, do not know each other, extra time is necessary for introductions and developing working relationships. If needed, information about the different team members should be distributed to all team members in advance of the orientation. Because of the importance of semistructured interviews for getting respondents to tell stories as opposed to just answering questions, the orientation may need to include practice interviewing, especially team interviewing. All team members need to understand that the purpose of the interview/topic guidelines is to provide structure for discussions and that all of the questions do not have to be asked of the respondents. If language interpreters are going to be used, it is critical that the team has experience with the use of the interpreters and that the interpreters understand the importance of reporting responses fully. The orientation should give as much attention to the iterative analysis/additional data collection process as to techniques for data collection. The orientation should include the use of the data display technique as part of the data analysis process, since this may be new to some team members (see appendix C on RQI training).

Logistics—Keeping RQI from Becoming SAP

> Attention to logistics helps keep RQI from becoming Slow Assessment Process (SAP).

Logistics and scheduling are the two critical factors for preventing Rapid Qualitative Inquiry (RQI) from turning into Slow Assessment Process (SAP). If the RQI is to be rapid, the logistics must be in place and must function. The use of computers and printers overseas can be especially problematic. The voltage may be different, electrical plugs will not fit, versions of operating systems and application programs may be different, paper sizes are likely to be different (A4 instead of letter), and there may be a need for software plug-ins to allow the printing of special characters. These problems can also be time consuming anywhere.

Assuming interviews are going to be recorded, someone should be responsible for testing the equipment and systems for saving files in multiple locations (see chapter 7). Transcribing tapes and preparing research logs can be terribly time consuming and a backlog in transcribing files can delay the entire process. Problems with transcription of tapes during the community college RQI illustrate this.

My experience has convinced me that (1) the team needs to have control over critical logistical issues such as interview transcription and (2) that having a person responsible for logistics is more important than having an additional academic discipline on the team (see ch. 5, "Team Leadership").

Schedule Flexibility

> Rapid does not mean rushed.

Rapid does not mean rushed. Schedules must be designed with sufficient flexibility that the team can take full advantage of unanticipated opportunities. Where a second appointment is scheduled soon after an earlier one, it may be useful to discuss flexibility with the second appointment and to have an agreed-upon way of making contact. If the team finds itself with a few minutes of unanticipated extra time, this time could be used for a moment of reflection or performing some analysis as part of the iterative process before the next appointment.

> The number of interviews may have to be limited.

One of the most difficult aspects of scheduling is planning for the number of interviews to be conducted. This is further complicated if interviews are to be done at different sites and the number of sites has to be considered. One of the characteristics of qualitative research is that it is impossible to know in advance exactly how many interviews will be needed. Part of the planning process is estimating the number of interviews and then remaining flexible based on the information that is collected. The community college RQI illustrates some of the problems associated with trying to interview more individuals than needed.

One approach for scheduling interviews is to (1) begin with a target number of interviews, (2) use more purposeful selection of the initial individuals to be interviewed, (3) interview others who want to participate on a first-come basis up to the target number of interviews, and (4) review and where appropriate revise the target number based on the results (see "Structuring the Research Time" and "How Much Data Is Needed?").

Improving Rigor while Collecting Data

Interview Notes in Addition to Recording

Even the best audio recorder will miss things that a good note taker will catch. It may be issues of context or expression, or a specific and possibly important word. Audio recorders fail to pick up words, or even entire sentences, or may fail all together. The MEMOS made by a note taker also can jump-start the analysis process. As noted earlier, the availability of the handwritten notes on the interviews during the community college RAP/RQI allowed the data analysis process to proceed even when there was a delay in the completion of transcripts of the tapes. Notes on interviews are especially useful when a language interpreter is used (see ch. 3, "Use of an Audio Recorder" and "Use of Interpreters").

> Audio recording technology should be expected to fail.

The Twenty-Four-Hour Rule

The twenty-four-hour rule is that research logs should be prepared within twenty-four hours of the observations. This means that recorded interviews should be transcribed and field notes typed within twenty-four hours. The resulting research logs also need to be reviewed, even if only casually, before the next cycle of data collection starts. The twenty-four-hour rule is often impossible to follow, but still it is important to try. Even traditional ethnographers who work alone note the importance of the preparation of the log within twenty-four hours to prevent the failure to capture, or to capture accurately, interviews and observations that were missed in the field notes or the interview transcript. For RQI, the twenty-four-hour rule is critical because of the need to explicitly consider the results of a day of data collection before beginning the next round (see ch. 3, "Managing Data—Field Notes and Logs").

Follow-Up Interviews with the Same Person

Traditional qualitative researchers and especially ethnographers who spend prolonged time in the field naturally return to the same individual numerous times. This is especially true of key informants who are able to discuss issues beyond their own experience. The iterative process of data collection and analysis helps to focus each subsequent meeting. The time constraints of RQI may make follow-up interviews with the same person appear to be a luxury. My experience has been that such interviews are extremely productive and sometimes may be more useful than an interview with a new person (see ch. 3, "Selection of Respondents").

> Follow-up interviews with the same person can sometimes be more useful than interviews with new respondents.

Spending Time at the Site

Eating and, when appropriate, sleeping at the site where an RQI is being done provides opportunities for informal discussions, follow-up discussions on topics of interest, and an understanding of the context of situations that may not be available otherwise. What comes up in the evening, at night, and in the early morning may never come up in interviews during "office hours." Even if it is not possible to share meals, it may be possible to share coffee or tea. The possible inconvenience for your hosts must always be considered when making a decision to spend extra time at the site. Chambers (2007) noted the importance of having unplanned time to "walk and wander around." Even students doing an RQI/Mini-RAP for a course should try to share a snack and a drink with their respondents.

Improving Rigor while Analyzing Data

Data Displays

The RQI team should experiment with a variety of ways for staying involved in the data analysis process from the beginning of the process. Data displays can be useful and ensure that all team members are involved in the process. Flip charts with bold color markers can be effective and are easily transported if the team is on the move. Tables based on comparisons between different data sources and drawings similar to rich pictures are good beginning points for data displays. Data displays should always be dated (see ch. 4, "Data Display," and ch.7, "Collaboration for Data Collection and Triangulation").

Presentation as a Way of Checking Back with Informants

Good qualitative inquiry is made better by checking back with the people who have provided the information. RQI places a special premium on checking as a part of the ongoing data analysis process. Presentation to the local community should be arranged at several points during the RQI and not just near the end. Early presentation, before the team becomes convinced of its findings, provides a real opportunity for input from the groups working with the RQI team. Preparation for presentations also is a way of ensuring that the data analysis process is ongoing (see ch. 4, "Checking Back with Informants").

Completion of the Draft Report

The RQI team, working together, should finish the draft report before the team disbands, even if this means less time collecting data. The preparation of the report is a critical part of the intensive team effort at data analysis. It usually makes sense to assign individual responsibility for the drafts of specific sections of the report. The sharing of these drafts provides wonderful opportunities for a team approach to data analysis. Word-processing programs that allow changes to be clearly marked or tracked and ways of sharing files such as the use of e-mail attachments, cloud-based digital tools, and other file sharing technology (see chapter 7) provide for collaboration on drafts without all parties having to be physically present at the same time. It is important for every team member to know what the other members have added or changed. Team members can check and comment on whether new additional quotes support conclusions that have already been reached and whether new or revised conclusions are appropriate. All team members can have the opportunity to review and edit as many drafts of a document as may be necessary (see ch. 4, "RQI Report Preparation by the Team").

Improving RQI Skills

Learning from Experience

Learning from experience while doing RQI is not automatic. Without explicit attention to lessons that can be learned, mistakes are likely to be repeated. Part of the ongoing data analysis process should be asking which techniques are working and which ones need to be modified. This same process should be done at the end of the RQI. A beginning question should always be whether RQI was the most appropriate research approach for the specific investigation. Specific questions should be asked about team membership and team interaction. A final question on the research method is what is too rapid or not rapid enough for a particular RQI.

> Either learn from experience or forever repeat the same mistakes.

Including Process as Well as Content in the Report

Most RQI reports focus on what was learned and may include some attention to the methods used. They do not usually reflect on process, such as who took part, what they did, or how they did it (Leurs 1997, 291). There are at least two reasons why reports should also pay some attention to process. The consumer of the report needs information on the process to determine how much confidence to place in the results. Attention to process is also necessary for the team to ensure that the process is as rigorous as it should be and that important elements have not been ignored. The RAP Sheet provides a beginning for a consideration of process, but is not adequate by itself.

Networking with Others about RQI

Other users of RQI are the best source of information about what works and what does not work. Technology, including blogs, and social media such as Facebook are excellent ways of networking with other users of RQI. The "QUALRS-L" listserver invites discussion of qualitative research methods in general, including discussions of rapid methods. Discussions of qualitative research, including rapid research methods, can be found on several web pages (see the introduction, "Online Resources," and ch. 7 on technology).

Sharing RQI Reports

Practitioners who use RQI and related rapid research methods are encouraged to publish their results in appropriate discipline-based journals. A list of journals that have published the results of rapid research methods is included in table 6.1. Practitioners of RQI are also encouraged to send information about publications or copies of their reports as e-mail attachments to beebe@gonzaga.edu. The goal is to make all reports available from the RQI homepage at http://rapidqualitativeinquiry.com.

Time Off

Even the most compulsive team member needs some time off. Team members need time away from each other, especially from the compulsive members. Weekends need to be honored, unless they are needed for data collection. In this case, another day should be identified for some time off. Experience with Rapid Appraisal in rural areas at Khon Kaen University in Thailand suggests that more than five hours per day spent in semi-structured interviewing and observation sessions is exhausting to even the heartiest team members and makes subsequent interviews less productive. More than about five days of this kind of fieldwork without a break can be counterproductive (Grandstaff and Grandstaff 1987, 78).

The Future of RQI and Some Concluding Comments Concerning Trends

As early as 1992 at a meeting on the use of Rapid Assessment Programs (RAP) with environmental programs, Scrimshaw observed that the accelerating rate of change in the world and a lessening of financial support for long-term research had driven the increasing interest in rapid research methods. According to Scrimshaw, the common thread for the evolution of rapid research methods was that "people reached the conclusion that the world was moving too quickly for the normal, disciplinary approaches." Scrimshaw also declared that "RAP is an idea whose time has come" (as cited in Abate 1992, 486).

More than twenty years after Scrimshaw made her comments, rapid research methods can be found in a growing number of fields and it is safe to assume that rapid research methods are here to stay. I firmly believe that rapid research methods such as RQI are relevant to a wide range of situations in which a qualitative approach to inquiry is needed. However, their use is limited by a failure to recognize the value of qualitative research in general and rapid qualitative research in particular. The challenge will be to embrace the potential of RQI, while at the same time recognizing its limitations. The major challenge to RQI is likely to continue to be confusing rapid with rushed. RQI is an idea whose time has come, but only if it is not oversold and only if it is implemented rigorously.

> RQI is an idea whose time has come?

The future of RQI will be closely connected with the future of qualitative research in general. Preissle (2011) provided an excellent place to begin a consideration of the future of RQI in her personal and very insightful discussion of the future of qualitative research in her article, "Qualitative Futures: Where We Might Go from Where We've Been."

I believe that rapid research such as RQI will likely be impacted by several specific trends including the following six.

1. Not only will RQI type methods be used globally, especially in but not limited to developing countries, but the assumptions for inquiry will be explicitly impacted by non-American leadership of this research. Increased attention will be given to the implications of differences in assumptions about power relationships resulting from colonialism (see Denzin and Giardina 2013). There will be a growing recognition of the need to decolonize research (see Smith 2012).

2. There will be less focus and debate about boundaries concerning research approaches, and techniques will be chosen to address specific issues while taking advantage of resources that are available. Greater flexibility in implementation is one of the most significant changes between traditional Rapid Rural Appraisal and Rapid Qualitative Inquiry. Flexibility can be expected to accelerate as individuals with different backgrounds become involved. In a blog on the future of qualitative research, one of the participants wrote of the need for a blurring of paradigms as a way of strengthening qualitative research (Social Science Seminar 2013).

3. There will be more focus on including the evidence that justified the conclusions in reports based on rapid research (see Tracy 2013). The American Education Research Association (AERA) (2006) identified specific standards for reporting on social science research (32–40) with specific recommendations for qualitative methods (37–38). Denzin (2011) summarized these recommendations: "Reports of research should be warranted, that is, adequate evidence, which would be credible (internal validity), should be provided to justify conclusions. Reports should be transparent, making explicit the logic of inquiry used in the project. This method should produce data that have external validity; reliability; and confirmability, or objectivity" (650).

4. There will be increased focus on variability with recognition that even small groups are not homogeneous. RQI uses purposeful selection of respondents and the differences among respondents can provide more insight than attempts to make statements about the average.

5. There will be increased attention on the context, including political and cultural, within which RQIs are implemented. There will also be increased attention to the implications of the context of inquiry for conclusions (see Creswell 2013).

6. The use of rapid research such as RQI will expand into new areas. The "epidemic" of the use of rapid research in the health field is likely to continue along with growing use in international development, business, cross-cultural awareness, and security. Other areas of potential growth for RQI are where an insider's perspective is especially important. Such areas include computer use, technology, communication, social media, policy, social work, organizational change, and leadership studies.

I look forward to working with others on making Rapid Qualitative Inquiry a tool for collaborative efforts to bring about change. I invite you to contact me: beebe@gonzaga.edu.

APPENDIX A
EXECUTIVE SUMMARY, COMMUNITY COLLEGE RAP/RQI

Summary of the Results
March 17, 2000
James Beebe, Dale Abendroth, Nancy Chase, and Grace Leaf

Most of the comments of the participants were grouped into categories and identified as constraints to the ability of the people involved in the Student Services Division to do the best job possible. There was general consensus that performance was not always as good as it should have been, that most people truly wanted to do better, but that there were a range of interrelated factors that prevented this from happening. The six constraints were (a) communication; (b) physical space; (c) technology; (d) utilization of people's time, talents, and creativity; (e) increases in the number and complexity of regulations; and (f) inadequate resources.

Communication

Virtually all of the interviews included references to problems with communication. Communication was seen as a problem at all levels, individual to individual, individual to supervisor, individual to administration, department to department, department to administration, and division to division. The need for open and consistent communication was expressed by many individuals. One of the most insightful comments on inadequate communications was voiced by a participant who said, "It's just quietness and you just start to wonder what's going on when it's so quiet."

Physical Space

Physical space was another theme that emerged as a constraint to people in doing the best job possible. Individuals expressed frustration at having to work in crowded, unattractive areas that afforded little privacy. A strong case was made by several participants for private offices as opposed to cubicles. "I think we need individual offices and not cubicles for the confidentiality that we have to work with. Even though I have a cubicle, you can hear everything. Students don't want to talk about their life when other people can hear."

Technology

There was a rather widespread conviction that technology, rather than reducing work or making the existing work easier, had merely added "insult to injury." Technology was seen as creating additional work and further complicating an already complicated environment. It was viewed as simply another load on employees and as a seriously limiting factor for many individuals. There was also a concern that the use of technology reduced personal contact with students. There was a widespread perception that many students were better prepared for technology than some of those within the Student Services Division. The students interviewed by the RAP team reported that they did not have a problem with technology, but often did have a problem with the way the community college implemented it.

Utilization of People's Time, Talents, and Creativity

Several members of the community referred to the tremendous enthusiasm, talents, creativity, and motivation of people with whom they work. Often these comments were in the context of "There is a lot of creativity that is stifled here." One participant was emphatic that "the only way we're going to really compete and be a dynamic institution is to recognize every staff person, every faculty person has a little leadership role."

Regulations

Numerous participants in this study commented upon the constraints imposed by regulations at the district, state, and federal levels. Comments were often in the context of increases in these regulations and often included a note of resignation. "We have to abide by federal regulations."

The need for some regulations was recognized. One participant noted that "we are a bureaucracy and if there wasn't some order to it, you know it would be total chaos, but sometimes it is difficult to get things to happen in as timely a fashion as you wish you could." There was a reluctant admission that there could sometimes be an advantage to the delays imposed by regulations.

Inadequate Resources

The general consensus was that resources for student services were declining. "I think in the time that I have been here, all I have ever seen is less money." A minority view was that overall resources were adequate but there was a problem with allocation. Many of these comments suggested that the administration was responsible for individual units getting less resources than they needed.

Staffing was the resource issue most often raised. Almost everyone commented on increased work levels, increased demands, reduced "downtime," and staff shortages.

Suggestions for dealing with inadequate resources, including staff, fell into three categories: (a) increase resources, (b) reallocate resources from one unit to another (usually from some other unit to "my" unit), and (c) use resources in better, more creative ways.

Facilitating Change

Specific suggestions for facilitating change emerged from the conversations about changes necessary for people to do the best job possible. The overarching theme was the need to develop trust. Trust was consistently linked to stability within the leadership team, open and consistent communication both vertically and horizontally, and developing a knowledge and appreciation of others' roles within student services. These themes were often interwoven in participant's conversation: "The more we know about each other, the more we trust each other, and if we trust each other, we'll be able to work with each other." Other suggestions to enhance the division's work included sharing information about successful approaches used in other places and the creation of venues to discuss new ways of doing things.

Numerous participants were highly skeptical that the RAP, or any study, could have a significant impact on conditions in the division. Several participants indicated that it was their hope that the RAP could be a factor in improving communication and would thereby facilitate change. There was, however, recognition that some of the issues that must be addressed were fundamental and there would be no easy or quick solutions. Within this context there was caution about expecting too much from any specific activity and a call for increasing the "good faith" that most agreed needed to be expanded.

APPENDIX B
EXECUTIVE SUMMARY, POLISH STATE FARMS RAP/RQI

Management and Transformation to a Market Economy: A Rapid Appraisal of State Farms in the Koszalin Voivodship, Poland[1]
October 11, 1991
James Beebe, John Farrell, Kent Olson, and Terrie Kolodziej with Tomasz Adamczyk.

State farms in Poland occupy about 18 percent of the agricultural land, but are a significant part of the economy and employ approximately 500,000 workers. The goal of the government of Poland was to eventually privatize these farms. A significant number of state farms were failing before they could be privatized and changes in management were necessary to counter this trend.

During the two years before this study, directors of state farms had been confronted with severe problems resulting from the transformation of the Polish economy. For the first time directors had to deal with high inflation and interest rates, with prices that did not always cover production costs, and with a general lack of demand. This environment, complicated by unclear government agricultural policy, had brought into focus differences in the management skills of directors.

Radical economic transformation provided the context for a rapid appraisal of the management skills of state farm directors in Koszalin Voivodship. The purpose of this rapid appraisal was to assess management responses to the economic environment. The principal research method was semistructured interviews of individual respondents and groups. Information from more than 110 hours of interviews was combined with direct observations of ten state farms. The collection and analysis of the data was an iterative process.

Variability in the reactions of state farm directors ranged from no response to changes in organizational structure, crop and livestock production, financial management, levels of employment, and marketing. These responses were analyzed based on the directors' explanations of what had happened. These responses suggested that many state farm directors did not fully understand how a market operates. This lack of full understanding appeared to be the most serious in areas of (1) risk, (2) information, (3) production vs. profits, (4) price determination, and (5) financial management.

Risk. From their comments, it appeared that many state farm directors believed the government should provide an essentially risk-free environment. They did not recognize that in a market economy government can reduce risk but cannot eliminate it. They tended to view risk as always negative, and failed to recognize that risk may provide opportunities for increased profits and growth for more creative managers.

Information. Many directors apparently had failed to recognize the power of information to reduce risk. They did not use available information as efficiently as they could have as a planning tool.

Production vs. profit. For many directors of state farms, decisions were aimed at achieving production goals. It will take considerable effort to convince state farm directors, bankers, and others that, under some conditions, levels of inputs will have to be cut and that physical production will and should fall.

Price determination. Many directors were not ready to accept that, in a market economy, prices are the result of supply and demand. Some directors found it incomprehensible that the government would not establish prices to cover their costs of production. Directors need to understand that in a market there may be times when prices will not cover costs.

Financial management. Many directors did not understand that good financial management is crucial in a market environment. In the past, state farm directors had access to subsidized short-term credit. With the reduction of credit availability, cash-flow projections and the implications of cash flow for the timing of activities became the most critical financial management skill for many state farm directors. There was general agreement that some of the most serious problems facing state farms cannot be resolved at the farm level. These included general agricultural policy, the monopoly position of farm output purchasers, bank policy, and the social role of state farms, especially the issue of apartments to the workers. The responses of the state farm directors, their understanding of how a market economy operates, and the problems that could not be resolved at the farm level led to three specific suggestions about changes in the advisory/extension services in Poland.

Focus on relative success. Because most directors appeared to recognize only a subset of the possible responses, the recommendation was that the advisory service disseminate information on the relatively more successful directors. The assumption was that this would help other directors recognize new options and convince them that success is possible.

Information. The second recommendation was that the agricultural advisory services identify specific information that directors and farmers want and the form in which it will be most useful and then make the information available.

Better understanding of the five issues related to a market economy. The third recommendation was that the advisory service help directors and others address the issues of risk, information, profits, price determination, and financial management by: (1) helping them locate and interpret market information necessary to predict a range of likely prices, (2) helping them adjust production decisions in response to this range of prices, and (3) helping them improve financial management, beginning with cash-flow analysis, in order to implement decisions based on price estimates.

An analysis of the situation faced by the directors of the state farms and the current role of the advisory services led to three specific suggestions for the advisory services: (1) Intensive training for members of the advisory services, the current directors of state farms, and future directors of these farms after privatization should be based on simulation, case studies, and the manipulation of "real-world" data (as opposed to classroom lectures). (2) The role of private sector Polish consultants should be encouraged and supported to complement the role of the public sector advisory service. (3) During the transformation process, the advisory service should assist in (a) valuing assets and dealing with legal issues, (b) identifying land that should not be in production and advising how to maintain and improve the quality of this land, and (c) promoting community development activities for unemployed state farm workers.

Note

1. Report prepared for the Extension Service of the United States Department of Agriculture (USDA) and the Polish Ministry of Agriculture and Food Economy under the Polish/American Extension Project.

APPENDIX C
RQI TRAINING FOR PRACTITIONERS AND STUDENTS

Both practitioners working on projects and students working on classroom activities based on doing short-term qualitative research may need RQI training. Rigorous implementation requires providing users with opportunities for developing appropriate skills and attitude.

> Training in appropriate skills and attitudes are needed to make RQI rapid but to prevent it from becoming quick and dirty.

Brief Introduction to Teaching Qualitative Research

As noted earlier, many of the skills and attitudes needed for rigorous implementation of RQI are similar to those needed for traditional qualitative research and developing these skills and attitudes is also similar.

There are two issues that make it difficult to learn to do qualitative research. As pointed out by Sherman and Webb in their 1988 article, the first difficulty is that the extremely complex nature of the discipline makes it difficult to have a definitive text. The second difficulty in teaching qualitative research is that the term *qualitative research* is used in confusing and ambiguous ways without clearly bounded content areas to be taught or specific procedures to follow (Hurworth 2008, 7).

Wheeler and Mallory (1996, 4) identified the choice for teaching qualitative research as between (a) field-based experiences and (b) reading examples of research. They noted that there appears to be consensus that one cannot teach qualitative research "simply by describing it, nor can students learn by only hearing a lecture or reading a textbook on the topic." They noted that reading examples of qualitative research can give students a feel for what is meant by qualitative research, but warned that only reading about qualitative research "may leave students in the dark or, worse, unconvinced" (5).

During a panel presentation at AERA Wolcott (1997 as cited by Hurworth 2008, 131) observed that over time his views on the choice for teaching qualitative research between "reading" and "doing" had shifted toward doing and he had developed a strong belief in "knowing how" rather than knowing "about." I especially appreciate as a summary on the importance of both doing and reading about doing qualitative research Michrina and Richards's statement (as cited by Hurworth 2008, 132) that "It is our belief that method without theory is indefensible and that theory without method is abstract and aimless. The two belong together and need to be taught together."

Hurworth's 2008 book, *Teaching Qualitative Research: Cases and Issues* is a good beginning place for exploring the issue of teaching qualitative research. Some of the key points made by Hurworth (2008) about teaching qualitative research include: the need for a combination of practice and introduction to theory behind the practice; practice in a group; and incremental learning, with time for review. These points are equally relevant for helping RQI users develop the skills and attitudes needed for RQI.

Learning to Implement RQI

For individuals who have had limited experience with qualitative techniques, there is a need to provide a strong rationale for and an introduction to qualitative research. For individuals with a background in qualitative research including students who have completed upper-division or graduate-level courses in qualitative research, anthropology, or sociology, there is a need to help them understand ways in which RQI differs from traditional qualitative research.

There is general consensus from users that RQI is best learned while participating as a team member with someone with experience, but that, since rapid research methods are "organized common sense," they largely can be self-taught. By reading reports by others, one can learn a significant amount about the methodology and understand realistic expectations for it.

Formal training often includes (a) orientations, (b) practice team interviewing, and (c) participation in a scaled-down version of an RQI called a "Mini-RAP." A Mini-RAP is not research, but an education activity.

> A Mini-RAP is not a research technique, but is an education activity for practicing skills and developing attitudes for doing RQI.

Orientation

In most cases, everyone who participates on an RQI team should attend an orientation. The online PowerPoint presentation on RQI can be a good beginning point for an orientation. The basic PowerPoint presentation can be found at http://rapidqualitativeinquiry.com. The basic PowerPoint will need to be modified based on the purpose of the planned RQI. Participants in the orientation should be reminded that in most cases what is presented is only one of a variety of ways of achieving the goal of RQI.

Orientation sessions can include practice semistructured solo interviews, practice team interviews, and practice team analysis. This book and handouts from the website can be provided as resources.

Practice Team Interview and Group Analysis

Even individuals who have had extensive experience with interviewing led by one person will probably need experience doing a team interview. Practice team interviews require teams of at least two persons doing at least one group interview of about ten minutes and ideally two group interviews. There is a need for time to prepare illustrative interview topics and questions, time after each interview to review both the content and the approach with a focus on what should change during the next team interview. Participants need to be reminded that analysis begins with the preparation of the illustrative interview topics and questions, and analysis intensifies after the collection of the first data.

> Team-based practice interviews need to include time for the team to review both what was learned and ways to improve the process.

Practice team interviews are also an opportunity to review the importance of conversational discussion as opposed to trying to get answers, and the difference between sequential interviewing based on team members taking turns asking questions and a shared approach that includes communication among the RQI team members. A topic that has worked well for practice team interviews is the meaning of gender equity. Time set aside to review the results of each interview with attention to lessons learned is as important as team members sharing the task of directing the discussion.

Someone with no experience with interviewing may need to do a short individual semistructured interview of five minutes or less. The practice interview should be videotaped. Just watching the results can have a significant impact on the time a researcher provides for others to respond before talking again. After watching the interview, the interviewer is asked to reflect on the assumptions she started with and how that influenced what was heard. The person interviewed is asked to report on how they understood the questions and other prompts. This is also an opportunity to consider whose categories are being used in the discussion. Groups of three provide a structure that allows everyone to lead the interview, be a respondent, and be an observer, critic, timekeeper, and the operator of the video equipment. It is important that video equipment and playback equipment are set up and easy to use. Topics should probably be assigned. One topic that seems to work in many situations is inquiry about choices of a college program or about future career plans. It may be necessary to remind everyone that the goal is to get the insider's perspective and to get stories as opposed to answers to questions. The need for informed consent needs to be discussed. There is also a need to remind people that even a five-minute interview should not be rushed.

> Watching a video of themselves leading a short practice semistructured interview can help researchers improve their sensitivity to providing respondents time to think.

Mini-RAP

A Mini-RAP can be the most important educational activity to prepare practitioners for doing an RQI or to prepare students for doing qualitative research. A Mini-RAP can help both practitioners and students develop the skills and attitudes needed for qualitative research. The activity is not called a Mini-RQI to avoid any suggestion that the abbreviated activity is a research activity or a short RQI. The Mini-RAP is no more like a regular RQI than a five-minute practice interview is like a regular qualitative interview.

At a minimum, a Mini-RAP is based on teams of at least two persons, two interviews of at least fifteen minutes, time to prepare illustrative interview topics and questions, time between the interviews to review both the content and the approach, and preparation of preliminary and tentative "findings." Participants in the activity should be expected to prepare interview transcripts, logs with coding, data analysis diagrams, and preliminary, tentative "findings."

Data collection for a Mini-RAP is significantly less than for an RQI and there is usually no insider on a Mini-RAP team. A Mini-RAP requires a minimum of two cycles of data collection with each interview a minimum of fifteen minutes long. When used as part of a qualitative research course, the process can be spread out over several weeks. Students can be given a choice of writing up the results in an "academic article" format as a class requirement or writing individual papers based on the results of their research.

Participants in a Mini-RAP are given the opportunity to begin practicing some of the specifics skills associated with qualitative research while working in a group. Specific skills include (a) team interviewing, with special attention to getting respondents to tell stories as opposed to answering questions; (b) producing transcripts and logs with MEMOS; (c) developing and implementing a coding system; (d) applying strategies for making sense out of results, including but not limited to use of data diagrams; and (e) drafting results.

Teamwork allows students to mentor each other. During interviews students quickly recognize good models and areas for improvements. Transcribing interviews takes time and in a group this task can be shared. Coding is always difficult the first time and at a minimum the frustration can be shared in a group. Wheeler and Mallory (1996, 9) contrasted the "blank faces or puzzled expressions" when individual students are asked to find themes and code data with a group effort where "they can address their own and each others' imposed values."

Training of Trainers

Training programs for RQI participants should be conducted by individuals who have had special training. A training of trainers program requires a minimum of about two weeks, would be based on this book supplemented by other references, and would include doing a regular RQI of at least four days.

The Center for Rapid Qualitative Inquiry (http://rapidqualitative inquiry.com) can certify RQI trainers.

Additional Readings

Hurworth's 2008 book, *Teaching Qualitative Research* is a comprehensive introduction to issues based on a review of the literature and qualitative research on training program. The article by Eisenhart and Jurow (2011), "Teaching Qualitative Research" in the *Sage Handbook*, is organized around fundamental questions involved in teaching a course and the implications of pedagogical approaches, structural constraints of where a course fits within a curriculum, and the debate in qualitative research between those who take a more conventional social science approach and focus on methods and those who take a more critical approach and focus on epistemological and ontological issues.

Eisenhart, Margaret, and A. Susan Jurow. 2011, Teaching qualitative research. In *The Sage handbook to qualitative research.* 4th ed., edited by Norman K. Denzin and Yvonna S. Lincoln, 699–714. Los Angeles, CA: Sage.

Hurworth, Rosalind. 2008. *Teaching qualitative research: Cases and issues.* Rotterdam, The Netherlands: Sense.

GLOSSARY

Appreciative Inquiry. Inquiry that focuses on identifying what is going well in a situation, determining the conditions that make excellence possible, and encouraging those conditions within the organizational culture (Hammond 2013).

bogus empowerment. Letting people think they have control over outcomes and the power to act on their own judgments when in reality they do not have this control or power. It can be the unintended consequence of hyperbole about the democratic nature of an organization (Ciulla 2003).

case study. A study involving efforts to understand a specific issue through one or more cases within a bounded system and based on in-depth data collection involving multiple sources of information (Creswell 2013).

coding. Part of the data analysis process consisting of the application of a limited number of labels to thought units (sentences, paragraphs, several paragraphs, or even individual words). Coding can be thought of as cutting the transcript of an interview into strips and placing the strips into piles. The labels for the piles are codes. Codes are based on threads that tie together bits of data. One usually starts with only five or six major codes and, if necessary, divides these codes into subcodes. Not all thought units are coded and some thought units can have multiple codes.

culture. Traditionally used by social scientists to describe nearly everything that has been learned or produced by a group of people. A more limited definition restricts the concept to the knowledge people gather, share, and use to generate and interpret social behavior (Spradley and McCurdy 1972).

data collection. The systematic gathering of facts, figures, and information from which conclusions can be inferred. For RQI, this includes information collected before interviews are conducted, what is observed, and what is heard. Semistructured interviews are a significant data collection technique for RQI.

data display. Part of the data analysis process for exploring the relationship between different pieces of information in which the goal is drawing conclusions. Data is usually displayed in the form of tables, charts, graphs, or drawings (Miles et al. 2014).

emic. Based on the categories the local people use for dividing up their reality and identifying the terms they use for these categories (Pelto and Pelto 1978).

etic. Based on the categories used by the researchers or outsiders to divide up reality (Pelto and Pelto 1978).

ethnography. A descriptive study of an intact cultural or social group or an individual or individuals within the group based primarily on participant observation and open-ended interviews. Reference to cultural terms helps differentiate ethnography from other qualitative research methods (Wolcott 2005). Ethnography is based on learning from people as opposed to studying people (Spradley and McCurdy 1972).

field notes. The usually handwritten notes that are done as the data is being collected. Transcripts of interviews that were recorded are sometimes called field notes.

fieldwork. The process of interacting and gathering information at the site or sites where a culture-sharing group is studied. The goal is to develop a portrait of a culture-sharing group.

individual respondents. Individuals who are interviewed about their own experience and not about their knowledge of the broader system. Questions should concern only the individual's knowledge and behavior, and not what he or she thinks about the knowledge and behavior of others. Individual respondents should be selected to represent variability.

informed consent. Permission provided by the participants in a study for the researcher to use the information participants provide. Permission should be based upon an understanding of the activity and the uses that will be made of the results. It should indicate that participation is voluntary and that the participant knows that he or she can withdraw at any time.

insiders. Team members or participants in social science research who are identified by the social group being studied as members in good standing. Traditional social science research has focused on efforts by investigators who have traditionally been outsiders trying to understand what the insiders believe, value, practice, and expect. There is a growing realization that insiders should be a part of the design, implementation, and publication of research (Van Maanen et al. 1996).

iterative process. A recursive process in which replications of a cycle produce results that approximate the desired result more and more closely. For RQI, the cycle of data analysis and data collection designed to produce better and better results is an iterative process.

key informants. Persons interviewed about the broader system and selected because of their experience and knowledge.

local community. The individuals in a culture-sharing group who are the subject of the RQI. Members of a local community do not need to be in one physical location. Members of a local community are sometimes identified as "local people" (see "insiders").

log. Logs are written documents that are the repositories of information from the field notes and transcripts, recorded in a format ready for analysis. Logs are a chronological record of what the team learns and the team's insights, and will include MEMOS. Logs are usually most useful if typed and double spaced, with very large margins on both sides. Lines may be numbered. They are usually prepared at a location away from where the observations were collected. Field notes and memory can be used to fill in missing words from transcripts to make logs more accurate.

margin remarks. Remarks written into the margin of the log, often the right-hand margin, after it has been typed. Almost anything can be included in a margin remark, but often these remarks are related to the coding activity. Some margin remarks suggest new interpretations and connections with other parts of the data and identify themes involving several different codes.

member checking. Before the conclusions are final, the RQI team should set aside time to share conclusions and to check facts with the people who have provided the information. Member checking provides an opportunity for local people to suggest their interpretations.

MEMOS. Reflections of the researcher, often notes to oneself about what has occurred, what has been learned, insights, and leads for future action. Also called "observer comments," "analytic memos," or "researcher memos." It is recommended that in an RQI log, MEMO be spelled with all uppercase letters and that their contents be included in square brackets "[ ]" to remind the team of the importance of not confusing reflections with observations (things heard or observed). MEMOS should be dated and, for an RQI, their author should be identified.

metaphor. A figure of speech in which a word or phrase, literally denoting one kind of object or idea, is used in place of another to suggest a likeness or analogy between them.

Mini-RAP. An educational activity based on RQI requiring a team of two people that conducts at least two interviews of about fifteen minutes each. A Mini-RAP includes attention to informed consent, development of illustrative topics, iterative data collection, data analysis, and additional data collection, identification of codes and themes, the identification of tentative findings, and a very brief report. A Mini-RAP is not an approach to research but is designed to provide an opportunity for developing skills and attitudes.

participant observation. A qualitative research technique based on becoming a part of the groups being studied and that requires intensive observing, listening, and speaking. The researcher participates in events as opposed to simply being there and passively watching (Ely 1991).

participants. Persons interviewed as part of the RQI process. *Participants* can be used interchangeably with *informants* or *respondents* when these terms are not modified with the words *individual* or *key*. The term *subjects* is generally avoided.

Participatory Rural Appraisal (PRA). A rapid/relaxed approach to research designed to enable local people to share, embrace, and analyze their knowledge of life and to plan and act based on mapping, diagramming, and comparison. The assumption is that local people will do almost all of the investigation and analysis and will then share with the outsiders (Chambers 2008).

propriety. Using procedures that are ethical and fair to those involved and affected by the results of the inquiry.

qualitative research/qualitative inquiry. Attempts to understand experience as the participants feel it and to interpret phenomena in terms of the meaning people bring to them. Studies are conducted in their natural settings. The researcher is an instrument of data collection (Miles et al. 2014). Qualitative research is based on a few cases and many variables, whereas quantitative research works with a few variables and many cases.

questionnaire/survey research. Research based on questionnaires, a group of written questions to which individuals respond. Questionnaires are often administered to a sample of subjects drawn from a population (a survey) where the objective is to be able to make inferences about the population. Even though questionnaire research and survey research are distinct methodologies, the terms are used interchangeably in this book to identify research in which specific questions are prepared in advance based on the assumption that local categories and words are known and that enough is known of the local situation to identify important issues.

Rapid Appraisal, Rapid Assessment, Rapid Rural Appraisal. Despite differences in details, there is general consensus that these terms describe research based on small multidisciplinary teams using semistructured interviews and direct observations to collect information, and that the entire process can be completed in one to six weeks.

Rapid Assessment Process (RAP). Intensive, team-based qualitative inquiry focused on the insiders' perspectives and multiple sources of data and triangulation, iterative data analysis, and additional data collection to quickly develop a preliminary understanding of a situation.

Rapid Qualitative Inquiry (RQI). Another term for Rapid Assessment Process that by definition is team based and has even more focus on the insiders' perspectives and on flexibility of implementation than RAP.

rapid research methods. Approaches to research designed to address the need for cost-effective and timely results in rapidly changing situations. These methods share many of the characteristics of qualitative research. The objective for most of these approaches can be summarized as first-cut assessments of poorly known areas (Conservation International 1991, 1).

reliability. The consistency of a measure from one use to the next. When repeated measurements of the same thing give similar results, the measurement is said to be reliable.

research tourism. An investigation that involves a quick visit to a study area and contact with the most easily reached local participants.

rich picture. A drawing by the participants in an activity and a way of encouraging especially less verbal participants to share their views. Rich pictures are an element of Soft Systems Methodology (Checkland and Poulter 2007).

semistructured interview. An interview using open-ended questions or probes designed to get local people to talk about a subject and not just answer specific questions. Guidelines for semistructured interviews should be viewed as a memory aid and not as an agenda to be diligently worked through. This method contrasts with the structured interview, in which identical questions are asked of every informant.

social science voyeurism. The practice by some social science researchers of observing or making inquiries on human behavior and practices to satisfy or gratify their own curiosity. Usually implies collecting more information than is necessary.

Soft Systems Methodology. An approach for rigorously considering messy and value-laden situations based on steps that include (1) identifying a situation which has provoked concern, (2) modeling the situation, (3) using the model to question the real-world situation, and (4) using the debate initiated by the comparison to define action that would improve the original problem situation (Checkland and Poulter 2007).

Sondeo. Term used by the Guatemalan Institute of Agricultural Science and Technology to describe a modified research technique for understanding the cropping or farming systems of farmers. Based on the use of multidisciplinary teams, semistructured interviews with farmers, and direct observation over a six- to ten-day period (Hildebrand 1979).

survey research. See "questionnaire/survey research."

tag-team interviewing. An unintended perception of the interview process by respondents who feel like they are being attacked by a gang of inquisitors trying to beat answers out of an uncooperative witness. A focus on getting respondents to tell their stories and a relaxed, conversational tone can prevent the perception of tag-team interviewing.

triangulation. Triangulation is used as a metaphor by social scientists for the use of data from different sources, the use of several different researchers, the use of multiple perspectives to interpret a single set of data, and the use of multiple methods to study a single problem. Triangulation does not imply there is a single truth.

validity. The degree to which research results reflect what the research claims, or purports, to be measuring. Even when measurements are reliable, consistent from one use to the next, they can be invalid. Some qualitative researchers propose the use of the term *trustworthiness* instead of *validity* and *reliability* (Ely 1991).

REFERENCES

Abate, Tom. 1992. Environmental rapid-assessment programs have appeal and critics. *Bioscience* 42 (7): 486–89.

AERA. 2006. Standards for reporting on empirical social science research in AERA publications: American Educational Research Association. *Educational Researcher* 35 (6): 33–40.

Agar, Michael H. 2013. *The lively science: Remodeling human social research*. Minneapolis, MN: Mill City Press.

———. 2004. Know when to hold 'em, know when to fold 'em: Qualitative thinking outside the university. *Qualitative Health Research* 14 (1): 100–112.

Angus, J., E. Hodnett, and L. O'Brien-Pallas. 2003. Implementing evidence-based nursing practice: A tale of two intrapartum nursing units. *Nursing Inquiry* 10 (4): 218–28.

Anker, M., R. J. Guidotti, S. Orzeszyna, S. A. Sapirie, and M. C. Thuriaux. 1993. Rapid evaluation methods (REM) of health services performance: Methodological observations. *Bulletin of the World Health Organization* 71 (1): 15–21.

Ante, Spencer E., and Cliff Edwards. 2006. The science of desire. *Business Week*, June 5, 98–106.

Ash, J. S., D. F. Sittig, K. P. Guappone, R. H. Dykstra, J. Richardson, A. Wright, J. Carpenter, C. McMullen, M. Shapiro, and A. Bunce. 2012. Recommended practices for computerized clinical decision support and knowledge management in community settings: A qualitative study. *BMC Medical Informatics and Decision Making* 12 (6). http://www.biomedcentral.com/1472–6947/12/6.

Auerswald, C. L., K. Greene, A. Minnis, I. Doherty, J. Ellen, and N. Padian. 2004. Qualitative assessment of venues for purposive sampling of hard-to-reach youth: An illustration in a Latino community. *Sexually Transmitted Diseases* 31 (2): 133–38.

Aylward, P., P. Murphy, K. Colmer, and M. O'Neill. 2010. Findings from an evaluation of an intervention targeting Australian parents of young children with attachment issues: The "Through the Looking Glass" (TtLG) project. *Australasian Journal of Early Childhood* 35 (3): 13–23.

Baldé, A. M. B. 2004. The schooling experiences of Fulani Muslim girls in the Fouta Djallon Region of Guinea: Forces influencing their retention in a rural secondary school of Dalaba. Dissertation. College of Education, Ohio University.

Bartsch, Dominik, and Nagette Belgacem. 2004. *Real time evaluation of UNHCR's response to the emergency in Chad.* Geneva, Switzerland: United Nations High Commissioner for Refugees Evaluation and Policy Unit, EPAU/2004/07.

Bartunek, Jean, and Meryl Reis Louis. 1996. *Insider/outsider team research.* Vol. 40, Qualitative Research Methods series. Thousand Oaks, CA: Sage.

Baxter, Pamela, and Susan Jack. 2008. Qualitative case study methodology: Study design and implementation for novice researchers. *Qualitative Report* 13 (4): 544–59.

Bazeley, Pat, and Kristi Jackson. 2013. *Qualitative data analysis with NVivo.* 2nd ed. Thousand Oaks, CA: Sage.

Becker, Howard Saul. 2007. *Writing for social scientists: How to start and finish your thesis, book, or article.* 2nd ed. Chicago Guides to Writing, Editing, and Publishing. Chicago: University of Chicago Press.

Bedford, J., M. Gandhi, M. Admassu, and A. Girma. 2012. "A normal delivery takes place at home": A qualitative study of the location of childbirth in rural Ethiopia. *Maternal and Child Health Journal*: 17 (2): 230–39. doi: 10.1007/s10995-012-0965-3.

Beebe, James. 2001. *Rapid assessment process: An introduction.* Walnut Creek, CA: AltaMira.

———. 1995. Basic concepts and techniques of rapid appraisal. *Human Organization* 54 (1): 42–51.

———. 1994. Concept of the average farmer and putting the farmer first. *Journal of Farming Systems Research-Extension* 4 (3): 1–16.

———. 1982. *Rapid rural appraisal, Umm Hijliij Breimya, el Obeid, Northern Kordolan.* Khartoum, Sudan: U.S. Agency for International Development.

Bentley, Margaret E., Gretel H. Pelto, Walter L. Straus, Debra A. Schumann, Catherine Adegbola, De La Pena, Gbolauan A. Oni, Kenneth H. Brown, and Sandra L. Huffman. 1988. Rapid ethnographic assessment: Applications in a diarrhea management program. *Social Science & Medicine* 27 (1): 107–16.

Bernard, H. Russell. 2011. *Research methods in anthropology: Qualitative and quantitative approaches*. 5th ed. Lanham, MD: AltaMira.

———. 1995. *Research methods in anthropology: Qualitative and quantitative approaches*. 2nd ed. Thousand Oaks, CA: Sage.

Blackburn, James, and Jeremy Holland. 1998. *Who changes? Institutionalizing participation in development*. London, UK: Intermediate Technology.

Bogdan, Robert, and Sari Knopp Biklen. 1998. *Qualitative research for education: An introduction to theory and methods*. 3rd ed. Boston: Allyn and Bacon.

Bonsa, S. 2003. The state of the private press in Ethiopia. In *Ethiopia, the challenge of democracy from below*, edited by Bahru Zewde and Singried Pausewang, 184–200. Stockholm, Sweden: Nordiska Afrikainstitutet. http:// nai.diva-portal.org/smash/get/diva2:242112/FULLTEXT01.pdf.

Bostain, J. C. 1970. *Suggestions for efficient use of an untrained interpreter*. Washington, DC: U.S. Department of State.

Bottrall, A. 1981. *Comparative study of the management and organization of irrigation projects*. Washington, DC: World Bank.

Bradley, R. T. 1982. Ethical problems in team research: A structural analysis and an agenda for resolution. *The Americans Sociologist* 17: 87–94.

BRIDGES (Briefings in Development and Gender). 1994. *Annotated bibliography on gender, rapid rural appraisal and participatory rural appraisal*. Brighton, UK: Institute of Development Studies, University of Sussex.

Brown, L. 2008. Using mobile learning to teach reading to ninth-grade students. Dissertation. Capella University.

Brown, M. S., M. Sebego, K. Mogobe, E. Ntsayagae, M. Sabone, and N. Seboni. 2008. Exploring the HIV/AIDS-related knowledge, attitudes, and behaviors of university students in Botswana. *Journal of Transcultural Nursing* 19 (4): 317–25.

Brugha, Ruairí, and Zsuzsa Varvasovszky. 2000. Stakeholder analysis: A review. *Health Policy and Planning* 15 (3): 239–46.

Bull, Susan, and Bobbie Farsides. 2012. Tailoring information provision and consent processes to research contexts: The value of rapid assessments. *Journal of Empirical Research on Human Research Ethics: JERHRE* 7 (1): 37–52.

Burgess, Robert G. 1982. *Field research: A sourcebook and field manual*. Contemporary Social Research Series, no. 4. Boston: G. Allen & Unwin.

California Postsecondary Education Commission. 2008. *Examining educational experiments: A field guide for conducting scientifically based research*. Sacramento: California Postsecondary Education Commission.

Caligiuri, P., R. Noe, R. Nolan, A. M. Ryan, and F. Drasgow. 2011. *Training, developing, and assessing cross-cultural competence in military personnel*. Technical

Report 1284. Arlington, VA: U.S. Army Research Institute for the Behavioral and Social Sciences. www.dtic.mil/dtic/tr/fulltext/u2/a559500.pdf.

Caracelli, V. J. 2006. Enhancing the policy process through the use of ethnography and other study frameworks: A mixed-method strategy. *Research in the Schools* 13 (1): 84–92.

Carley, Kathleen M., Jürgen Pfeffer, Jeff Reminga, Jon Storrick, and Dave Columbus. 2012. *ORA user's guide*. Report ADA563063. Fort Belvoir, VA: Defense Technical Information Center. http://www.dtic.mil/docs/citations/ADA563063.

CASL (Community Adaptation and Sustainable Livelihoods). 1999. *Participatory research for sustainable livelihoods: A guide for field projects on adaptive strategies: Participatory rural appraisal (PRA)*. http://www.iisd.org/casl/caslguide/pra.htm.

Chambers, Robert. 2008. *Revolutions in development inquiry*. London: Earthscan.

———. 2007. *Who counts? The quiet revolution of participation and numbers*. Brighton, UK: Institute of Development Studies, University of Sussex.

———. 1999. *Relaxed and participatory appraisal: Notes on practical approaches and methods*. Brighton, UK: Institute of Development Studies, University of Sussex.

———. 1997. *Whose reality counts? Putting the first last*. London: Intermediate Technology.

———. 1996. Introduction to participatory approaches and methodologies. Included in Chamber (1999) *Relaxed and participatory appraisal: Notes on practical approaches and methods*.

———. 1994. The origins and practice of participatory rural appraisal. *World Development* 22 (7): 953–69.

———. 1991. Shortcut and participatory methods for gaining social information for projects. In *Putting people first: Sociological variables in rural development*. 2nd ed., edited by M. M. Cernea, 515–37. Washington, DC: Oxford University Press, World Bank.

———. 1983. *Rapid appraisal for improving existing canal irrigation systems*. New Delhi, India: Ford Foundation.

———. 1980. Shortcut methods in information gathering for rural development projects. Paper presented at the World Bank Agricultural Sector Symposium. Brighton, UK: Institute of Development Studies, University of Sussex.

Chambers, Robert, and J. Blackburn. 1996. *The power of participation*. Brighton, UK: Institute of Development Studies, University of Sussex.

Checkland, Peter. 1981. *Systems thinking, systems practice*. Chichester, UK: Wiley.

Checkland, Peter, and John Poulter. 2007. *Learning for action: A short definitive account of soft systems methodology and its use for practitioner, teachers, and students*. Chichester, UK: Wiley.

Checkland, Peter, and Jim Scholes. 1999. *Soft systems methodology in action: A 30-year retrospective*. New ed. Chichester, UK: Wiley.

Ciulla, Joanne B. 2004. Leadership and the problem of bogus empowerment. In *Ethics: The heart of leadership*, edited by J. B. Ciulla, 59–82. Westport, CT: Praeger.

———. 2003. *The ethics of leadership*. Belmont, CA: Thomson/Wadsworth.

Clarke, Anthony, and Gaalen Erickson. 2003. *Teacher inquiry: Living the research in everyday practice*. New York: Routledge.

Clement, Ulrich. 1990. Surveys of heterosexual behaviour. *Annual Review of Sex Research* 1: 45–74.

Cohen, Susan G., and Diane E. Bailey. 1997. What makes teams work: Group effectiveness research from the shop floor to the executive suite. *Journal of Management* 23 (3): 239–99.

Conservation International. 1991. *A biological assessment of the Alto Madidi region and adjacent areas of northwest Bolivia: Rapid assessment program*. RAP Working Papers no. 1. Washington, DC: Conservation International.

Cooperrider, David L., and Diana Kaplin Whitney. 2005. *Appreciative inquiry: A positive revolution in change*. San Francisco, CA: Berrett-Koehler.

Cordingley, P. 2003. Research and evidence-based practice: Focusing on practice and practitioners. In *Developing educational leadership: Using evidence for policy and practice*, edited by Lesley Anderson and Nigel Bennett. London: Sage.

Creswell, John W. 2014. *Research design: Qualitative, quantitative, and mixed methods approaches*. 4th ed. Thousand Oaks, CA: Sage.

———. 2013. *Qualitative inquiry and research design: Choosing among five approaches*. 3rd ed. Los Angeles, CA: Sage.

———. 2008. *Educational research: Planning, conducting, and evaluating quantitative and qualitative research*. 3rd ed. Upper Saddle River, NJ: Pearson/Merrill Prentice Hall.

———. 2007. *Qualitative inquiry and research design: Choosing among five approaches*. 2nd ed. Thousand Oaks, CA: Sage.

———. 1998. *Qualitative inquiry and research design: Choosing among five traditions*. Thousand Oaks, CA: Sage.

Cullen, E. T., L. N. H. Matthews, and T. D. Teske. 2008. Use of occupational ethnography and social marketing strategies to develop a safety awareness campaign for coal miners. *Social Marketing Quarterly* 14 (4): 2–21.

Daack-Hirsch, S., and H. Gamboa. 2011. Working-class Filipino women's perspectives on factors that facilitate or hinder prenatal micronutrients supplementation to prevent congenital anomalies. *Asia-Pacific Journal of Public Health* 24 (6):1023–35. doi: 10.1177/1010539511406711.

Davidson, Judith, and Silvana di Gregorio. 2011. Qualitative research and technology: In the midst of a revolution. In *Handbook of qualitative research*. 4th ed., edited by Norman K. Denzin and Yvonna S. Lincoln, 627–43. Thousand Oaks, CA: Sage.

Denzin, Norman R. 2011. The politics of evidence. In *The Sage handbook of qualitative research*, edited by Norman K. Denzin and Yvonna S. Lincoln, 645–58. Los Angeles, CA: Sage.

———. 2009. *Qualitative inquiry under fire: Toward a new paradigm dialogue*. Walnut Creek, CA: West Coast Press.

Denzin, Norman K., and Michael D. Giardina. 2013. *Global dimensions of qualitative inquiry*. Walnut Creek, CA: Left Coast Press.

Denzin, Norman K., and Yvonna S. Lincoln. 2011. *The Sage handbook of qualitative research*. 4th ed. Thousand Oaks, CA: Sage.

———. 2005. *The Sage handbook of qualitative research*. 3rd ed. Thousand Oaks, CA: Sage.

———. 1994. *Handbook of qualitative research*. Thousand Oaks, CA: Sage.

Driscoll, D., A. Sorensen, and M. Deerhake. 2012. A multidisciplinary approach to promoting healthy subsistence fish consumption in culturally distinct communities. *Health Promotion Practice* 13 (2): 245–51.

Eisenhart, Margaret, and A. Susan Jurow. Teaching qualitative research. In *The Sage handbook of qualitative research*. 4th ed., edited by Norman K. Denzin and Yvonna S. Lincoln, 699–714. Los Angeles, CA: Sage.

Ely, Margot, with Margaret Anzul, Teri Friedman, Diane Garner, and Ann McCormack Steinmetz. 1991. *Doing qualitative research: Circles within circles*. New York: The Falmer Press.

Erickson, Ken C., and Donald D. Stull. 1998. *Doing team ethnography: Warnings and advice*. Vol. 42, Qualitative Research Methods Series. Thousand Oaks, CA: Sage.

Ervin, Alexander M. 2005. *Applied anthropology: Tools and perspectives for contemporary practice*. 2nd ed. Boston: Pearson/Allyn and Bacon.

Fernald, Douglas H., and Christine W. Duclos. 2005. Enhance your team-based qualitative research. *Annals of Family Medicine* 3 (4): 360–64.

Fetterman, David M. 2010. *Ethnography: Step-by-step*. Applied Social Research Methods series. 3rd ed. Los Angeles, CA: Sage.

———. 1998. *Ethnography: Step by step*. Applied Social Research Methods series. 2nd ed. Thousand Oaks, CA: Sage.

Fielding, Nigel, and Jane L. Fielding. 1986. *Linking data*. Vol. 4, Qualitative Research Methods. Beverly Hills, CA: Sage.

Finnie, R. K. C., T. Mabunda, L. B. Khoza, B. van den Borne, B. Selwyn, and P. D. Mullen. 2010. Pilot study to develop a rapid assessment of tuberculosis care-seeking and adherence practices in rural Limpopo province, South Africa. *International Quarterly of Community Health Education* 31 (1): 3–19.

Fitch, C., T. Rhodes, and G. V. Stimson. 2000. Origins of an epidemic: The methodological and political emergence of rapid assessment. *International Journal on Drug Policy* 11 (1–2): 63–82.

Flick, Uwe. 1992. Triangulation revisited: Strategy of validation or alternative? *Journal for the Theory of Social Behaviour* 22 (2): 175–97.

Fluehr-Lobban, Caroline. 1998. Ethics. In *Handbook of methods in cultural anthropology*, edited by H. Russell Bernard, 178–202. Walnut Creek, CA: AltaMira.

Freeman, R. Edward. 1984. *Strategic management: A stakeholder approach*. Pitman Series in Business and Public Policy. Boston: Pitman.

Friedemann-Sanchez, G., N. A. Sayer, and T. Pickett. 2008. Provider perspectives on rehabilitation of patients with polytrauma. *Archives of Physical Medicine and Rehabilitation* 89 (1): 171–78.

Galt, Daniel. 1987. How rapid rural appraisal and other socio-economic diagnostic techniques fit into the cyclic FSR/E process. Paper presented at Proceedings of the 1985 International Conference on Rapid Rural Appraisal, Khon Kaen University, Khon Kaen, Thailand.

Gifford, B., A. Kestler, and S. Anand. 2010. Building local legitimacy into corporate social responsibility: Gold mining firms in developing nations. *Journal of World Business* 45 (3): 304–11.

Gilden, P. B. Y. 2005. *Social science in the Pacific fishery management council process*. Portland, OR: Pacific Fishery Management Council.

Gow, David D. 1991. Collaboration in development consulting: Stooges, hired guns, or musketeers? *Human Organization* 50 (1): 1–15.

Grandstaff, T. B., and S. W. Grandstaff. 1987. A conceptual basis for methodological development in rapid rural appraisal. Paper presented at Proceedings of the 1985 International Conference on Rapid Rural Appraisal, Khon Kaen University, Khon Kaen, Thailand.

Greenhow, Christine, Beth Robelia, and Joan E. Hughes. 2009. Learning, teaching, and scholarship in a digital age: Web 2.0 and classroom research—what path should we take "now"? *Educational Researcher* 38 (4): 246–59.

Guerrero, M. L., R. C. Morrow, J. J. Calva, H. Ortega-Gallegos, S. C. Weller, G. M. Ruiz-Palacios, and A. L. Morrow. 1999. Rapid ethnographic assessment of breastfeeding practices in periurban Mexico City. *Bulletin of the World Health Organization: The International Journal of Public Health* 77 (4): 323–30.

Guest, Greg, and Kathleen M. MacQueen. 2008. *Handbook for team-based qualitative research*. Lanham, MD: AltaMira.

Guest, Greg, Emily E. Namey, and Marilyn L. Mitchell. 2013. *Collecting qualitative data: A field manual for applied research*. Thousand Oaks, CA: Sage.

Hammersley, Martyn, and Paul Atkinson. 1995. *Ethnography: Principles in practice.* 2nd ed. New York: Routledge.

Hammond, Sue Annis. 2013. *The thin book of appreciative inquiry.* 3rd ed. Plano, TX: Thin Books.

Handwerker, W. Penn. 2006. The evolution of ethnographic research methods: Curiosities and contradictions in the qualitative research literature. *Reviews in Anthropology* 35 (1): 105–18.

———. 2001. *Quick ethnography.* Walnut Creek, CA: AltaMira.

Harris, K. J., N. W. Jerome, and S. B. Fawcett. 1997. Rapid assessment procedures: A review and critique. *Human Organization* 56 (3): 375–78.

Heifetz, Ronald A. 1994. *Leadership without easy answers.* Cambridge, MA: Belknap Press of Harvard University.

Hildebrand, Peter E. 1979 Summary of the Sondeo methodology used by ICTA. Paper presented at the Rapid Rural Appraisal Conference at the Institute of Development Studies, University of Sussex, Brighton, England.

Honadle, G. 1979. *Rapid reconnaissance approaches to organizational analysis for development administration.* Washington, DC: Development Alternatives.

Hurworth, Rosalind E. 2008. *Teaching qualitative research: Cases and issues.* Rotterdam, The Netherlands: Sense.

Inciardi, J. A., H. L. Surratt, T. J. Cicero, and R. A. Beard. 2009. Prescription opioid abuse and diversion in an urban community: The results of an ultrarapid assessment. *Pain Medicine* 10 (3): 537–48.

Jamal, A., and J. Crisp. 2002. *Real-time humanitarian evaluations: Some frequently asked questions.* Geneva, Switzerland: Evaluation and Policy Analysis Unit, United Nations High Commissioner for Refugees.

Janesick, Valerie. 1994. The dance of qualitative research design: Metaphor, methodolatry, and meaning. In *Handbook of qualitative research*, edited by Norman K. Denzin and Yvonna S. Lincoln, 209–19. Thousand Oaks, CA: Sage.

Johnsnen, T. E., and P. Helmersen. 2009. Contextualizing customers. *Ethnographic Praxis in Industry Conference Proceedings 2000* (1): 185–96.

Keesing, Roger, and Andrew Strathern. 1998. Fieldwork. In *Cultural anthropology: A contemporary perspective.* 3rd ed., edited by Roger Keesing and Andrew Strathern, 7–10. Forth Worth, TX: Harcourt Brace.

Khon Kaen University. 1987. *Proceedings of the 1985 International Conference on Rapid Rural Appraisal: Khon Kaen University.* International Conference on Rapid Rural Appraisal. Khon Kaen, Thailand: Rural Systems Research Project and Farming Systems Research Project for Khon Kaen University.

Kresno, S., G. G. Harrison, B. Sutrisna, and A. Reingold. 1994. Acute respiratory illnesses in children under five years in Indramayu, West Java, Indonesia: A rapid ethnographic assessment. *Medical Anthropology* 15 (4): 425–34.

Krueger, Cornelia Corinna. 2006. The impact of the Internet on business model evolution within the news and music sectors. PhD Thesis. School of Computer and Information Science University of South Australia, Adelaide, Australia.

Krueger, Richard A., and Mary Anne Casey. 2009. *Focus groups: A practical guide for applied research.* 4th ed. Los Angeles, CA: Sage.

Kumar, K. 1993. *Rapid appraisal methods.* World Bank Regional and Sectoral Studies. Washington, DC: World Bank.

———. 1987. *Conducting group interviews in developing countries.* Washington, DC: U.S. Agency for International Development.

Kvale, Steinar, and Svend Brinkmann. 2009. *InterViews: Learning the craft of qualitative research interviewing.* 2nd [rev.] ed. Los Angeles, CA: Sage.

Lagacé, R. O. 1970. The HRAF data quality control schedule. *Behavior Science Notes* 5 (2): 125–32.

Lam, V. Y. Y., and Y. Sadovy de Mitcheson. 2011. The sharks of Southeast Asia: Unknown, unmonitored and unmanaged. *Fish and Fisheries* 12 (1): 51–74.

Lassiter, Luke E. 2005. *The Chicago guide to collaborative ethnography.* Chicago Guides to Writing, Editing, and Publishing. Chicago: University of Chicago Press.

Last, L. C. D. 2005. *Rapid assessment process (RAP) and security sector reform.* Kingston, Canada: Royal Military College.

Lee, Raymond M. 1994. *Dangerous fieldwork.* Vol. 34, Qualitative Research Methods series. Thousand Oaks, CA: Sage.

Leurs, Robert. 1997. Critical reflections on rapid and participatory rural appraisal. *Development in Practice* 7 (3): 291–93.

Lévy, Pierre. 2001. Collective intelligence. In *Reading in digital culture*, edited by David Trend, 253–58. Malden, MA: Blackwell.

———. 1997. *Collective intelligence: Mankind's emerging world in cyberspace.* Cambridge, MA: Perseus Books.

Lincoln, Yvonna S., and E. G. Guba. 1985. *Naturalistic inquiry.* Beverly Hills, CA: Sage.

Low, S. M., D. H. Taplin, and M. Lamb. 2005. Battery Park City: An ethnographic field study of the community impact of 9/11. *Urban Affairs Review* 40 (5): 655–82. doi: 10.1177/1078087404272304.

Macintyre, K. 1995. *The case for rapid assessment surveys for family planning program evaluation.* Chapel Hill: University of North Carolina.

MacQueen, Kathleen M., and Greg Guest. 2008. An introduction to team-based qualitative research. In *Handbook for team-based qualitative research*, edited by Greg Guest and Kathleen M. MacQueen, 3–20. Lanham, MD: AltaMira.

Marshall, Catherine, and Gretchen B. Rossman. 2011. *Designing qualitative research.* 5th ed. Los Angeles, CA: Sage.

Mathison, Sandra. 1988. Why triangulate? *Educational Researcher* 17 (2): 13–17.

McCurdy, David W., James P. Spradley, and Dianna J. Shandy. 2005. *The cultural experience: Ethnography in complex society*. 2nd ed. Long Grove, IL: Waveland Press.

McDonald, J. 2009. Bulldozers, land, and the bottom: Environmental justice and a rapid assessment process. *Practicing Anthropology* 31 (1): 4–8.

McNall, Miles, and Pennie Foster-Fishman. 2007. Methods of rapid evaluation, assessment, and appraisal. *American Journal of Evaluation* 28 (2): 151–68.

McNall, Miles A., Vincent E. Welch, Kari L. Ruh, Carolyn A. Mildner, and Tomas Soto. 2004. The use of rapid-feedback evaluation methods to improve the retention rates of an HIV/AIDS healthcare intervention. *Evaluation & Program Planning* 27 (3): 287–94.

Merriam, Sharan B. 1998. *Qualitative research and case study applications in education*. 2nd ed. San Francisco, CA: Jossey-Bass.

Metzler, Ken. 1997. *Creative interviewing: The writer's guide to gathering information by asking questions*. 3rd ed. Boston: Allyn and Bacon.

Mignone, J., G. M. Hiremath, V. Sabnis, J. Laxmi, S. Halli, J. O'Neil, B. M. Ramesh, J. Blanchard, and S. Moses. 2009. Use of rapid ethnographic methodology to develop a village-level rapid assessment tool predictive of HIV infection in rural India. *International Journal of Qualitative Methods* 8 (3): 68–83.

Miles, Matthew B., and A. Michael Huberman. 1994. *Qualitative data analysis: An expanded sourcebook*. 2nd ed. Thousand Oaks, CA: Sage.

Miles, Matthew B., A. Michael Huberman, and Johnny Saldaña. 2014. *Qualitative data analysis: A methods sourcebook*. 3rd ed. Thousand Oaks, CA: Sage.

Millen, D. R. 2000. Rapid ethnography: Time deepening strategies for HCI field research. *Proceedings of the 3rd Conference on Designing Interactive Systems: Processes, Practices, Methods, and Techniques*: 280–86. doi: 10.1145/347642.347763.

Miller, Tina, Maxine Birch, Melanie Mauthner, and Julie Jessop (Eds.). 2012. *Ethics in qualitative research*. 2nd ed. Los Angeles, CA: Sage.

Morgan, David L. 1997. *Focus groups as qualitative research*. Vol. 16, Qualitative Research Methods series. Los Angeles, CA: Sage.

Morin, S. F., K. A. Koester, W. T. Steward, A. Maiorana, M. McLaughlin, J. J. Myers, K. Vernon, and M. A. Chesney. 2004. Missed opportunities: Prevention with HIV-infected patients in clinical care settings. *Journal of Acquired Immune Deficiency Syndromes* 36 (4): 960–66.

Mpondi, D. 2004. Educational change and cultural politics: National identity-formation in Zimbabwe Dissertation. Ohio University.

Myers, Michael D. 2008. *Qualitative research in business and management*. Los Angeles, CA: Sage.

Neuwirth, E. E. B., J. A. Schmittdiel, K. Tallman, and J. Bellows. 2007. Understanding panel management: A comparative study of an emerging approach to population care. *The Permanente Journal* 11 (3): 12–20.

Nunns, Heather. 2009. Responding to the demand for quicker evaluation findings. *Social Policy Journal of New Zealand* 34: 89–99.

Oakes, J. Michael. 2002. Risks and wrongs in social science research: An evaluator's guide to the IRB. *Evaluation Review* 26 (5): 443–79.

Overseas Development Administration (ODA). 1995. *Guidance note on how to do a stakeholder analysis of aid projects and programmes.* London: Overseas Development Administration.

Patton, Michael Quinn. 2002. *Qualitative research and evaluation methods.* 3rd ed. Thousand Oaks, CA: Sage.

Paulus, Trena, Jessica Nina Lester, and Paul Dempster. 2014. *Digital tools for qualitative research.* Thousand Oaks, CA: Sage.

Pelto, Pertti J., and Gretel H. Pelto. 1978. *Anthropological research: The structure of inquiry.* 2nd ed. New York: Cambridge University Press.

Pennesi, Karen. 2007. Improving forecast communication: Linguistic and cultural considerations. *Bulletin of the American Meteorological Society* 88 (7): 1033–44.

Phillips, R. S. 1993. Geographic knowledge and survey research. *International Journal of Public Opinion Research* 5 (1): 100–104.

Pomeroy, Caroline, and Melissa M. Stevens. 2008. *Santa Cruz harbor commercial fishing community profile.* T–066. San Diego: University of California, California Sea Grant Extension Program.

Preissle, Judith. 2011. Qualitative futures: Where we might go from where we've been. In *The Sage handbook of qualitative research.* 4th ed., edited by Norman K. Denzin and Yvonna S. Lincoln. Los Angeles, CA: Sage.

Punch, Maurice. 1986. *The politics and ethics of fieldwork.* Vol. 3, Qualitative Research Methods. Beverly Hills, CA: Sage.

Reeves, E. B., and T. Frankenberger. 1981. *Socio-economic constraints to the production, distribution, and consumption of millet, sorghum, and cash crops in North Kordofan, Sudan.* Lexington: University Press of Kentucky.

Reid, Alan. 2004. *Towards a culture of inquiry in DECS.* South Australia: Department of Education and Children's Services.

Rhoades, R. E. 1987. Basic field techniques for rapid rural appraisal. In *Proceedings of the 1985 International Conference on Rapid Rural Appraisal*, ed. Khon Kaen University, 114–28. Khon Kaen, Thailand: Rural Systems Research and Farming Systems Research Projects.

———. 1982. *The art of the informal agricultural survey.* Lima, Peru: International Potato Center.

Richards, Lyn. 2009. *Handling qualitative data: A practical guide*. 2nd ed. London: Sage.

———. 1999. Qualitative teamwork: Making it work. *Qualitative Health Research* 9 (1): 7–10.

Richards, Lyn, and Janice M. Morse. 2013. *Readme first for a user's guide to qualitative methods*. 3rd ed. Los Angeles, CA: Sage.

Richardson, L. 1994. Writing: A method of inquiry. In *Handbook of qualitative research*, edited by Norman K. Denzin and Yvonna S. Lincoln, 516–29. Thousand Oaks, CA: Sage.

Rifkin, S. B. 1996. Rapid rural appraisal: Its use and value for health planners and managers. *Public Administration* 74 (Autumn): 509–26.

Robb, Caroline M. 2002. *Can the poor influence policy? Participatory poverty assessments in the developing world*. 2nd ed. Washington, DC: International Monetary Fund and the World Bank.

Saldaña, Johnny. 2013. *The coding manual for qualitative researchers*. 2nd ed. Los Angeles, CA: Sage.

———. 2011. *Fundamentals of qualitative research: Understanding qualitative research*. New York: Oxford University Press.

Sandhu, Jaspal, P. Altankhuyag, and D. Amarsaikhan. 2007. Serial hanging out: Rapid ethnographic needs assessment in rural settings. *Human-computer interaction: Interaction design and usability*. Part 1, 614–23. Berlin: Soringer Berlin Heidelberg. doi: 10.1007/978-3-540-73105-4_68.

Sandison, Peta. 2003. *Desk review of real-time evaluation experience*. New York: United Nations Children's Fund (UNICEF).

Savage, J. 2006. Ethnographic evidence the value of applied ethnography in healthcare. *Journal of Research in Nursing* 11 (5): 383–93.

Sayer, N. A., N. A. Rettmann, K. F. Carlson, N. Bernardy, B. J. Sigford, J. L. Hamblen, and M. J. Friedman. 2009. Veterans with history of mild traumatic brain injury and posttraumatic stress disorder: Challenges from provider perspective. *Journal of Rehabilitation Research & Development* 46 (6): 703–16.

Schneider, Jo Anne. 2013. *Qualitative research and IRB: A comprehensive guide for IRB forms, informed consent, writing IRB applications and more*. Bonita Springs, FL: Principal Investigators Association.

Schultz, Jack, Peter Van Arsdale, and Ed Knop. 2009. *Rapid ethnographic assessment final report*. Boulder, CO: eCrossCulture for SITIS, US Military, OSD08-CR3.

Scrimshaw, Nevin S., and Gary Richard Gleason. 1992. *RAP, rapid assessment procedures: Qualitative methodologies for planning and evaluation of health related programmes*. Boston: International Nutrition Foundation for Developing Countries.

Shaner, W. W., P. F. Philipp, and W. R. Schmehl. 1982. *Farming systems research and development: Guidelines for developing countries*. Boulder, CO: Westview Press.

Shenton, Andrew. 2004. Strategies of ensuring trustworthiness in qualitative research projects. *Education for Information* 22: 63–75.

Sherman, Robert R., and Rodman B. Webb. 1988. *Qualitative research in education: Focus and methods*. Explorations in Ethnography series. New York: The Falmer Press.

Siltanen, Janet, Alette Willis, and Willow Scobie. 2008. Separately together: Working reflexively as a team. *International Journal of Social Research Methodology* 11 (1): 45–61.

Silverman, George. 2011. How to get beneath the surface in focus groups. George Silverman's marketing strategy secrets. Focus Group Center. http://mnav.com/focus-group-center/bensurf-htm.

Smith, Linda Tuhiwai. 2012. *Decolonizing methodologies: Research and indigenous peoples*. 2nd ed. New York: Zed Books.

Social Sciences Seminar. 2013. What do we see as the future of qualitative research? Class Blog. April 8, 2013. http://mla507.wordpress.com/2013/04/08/what-do-we-see-as-the-future-of-qualitative-research.

Solomon, P. L., J. A. Tennille, D. Lipsitt, E. Plumb, D. Metzger, and M. B. Blank. 2007. Rapid assessment of existing HIV prevention programming in a community mental health center. *Journal of Prevention & Intervention in the Community* 33 (1–2): 137–51.

Sonnichsen, Richard C. 2000. *High impact internal evaluation: A practitioner's guide to evaluating and consulting inside organizations*. Thousand Oaks, CA: Sage.

Spoon, Jeremy, and Richard Arnold. 2012. Collaborative research and co-learning: Integrating Nuwuvi (Southern Paiute) ecological knowledge and spirituality to revitalize a fragmented land. *Journal for the Study of Religion, Nature, and Culture* 6 (4): 477–500.

Spradley, James P. 1980. *Participant observation*. New York: Holt, Rinehart and Winston.

———. 1979. *The ethnographic interview*. New York: Holt, Rinehart and Winston.

Spradley, James P., and David W. McCurdy. 1972. *The cultural experience: Ethnography in complex society*. Chicago: Science Research Associates.

Stake, Robert E. 2006. *Multiple case study analysis*. New York: Guilford Press.

———. 2005. Qualitative case studies. In *The Sage handbook of qualitative research*. 3rd ed., edited by Norman K. Denzin and Yvonna S. Lincoln, 443–66. Thousand Oaks, CA: Sage.

———. 1995. *The art of case study research*. Thousand Oaks, CA: Sage.

Stein, Eric, Mathaa Sutula, A. Elizabeth Fetscher, Ross P. Clark, Joshua N. Collins, J. Letitia Grenier, and Cristina Grosso. 2009. Diagnosing wetland health with rapid assessment methods, Report 2009–0010. *Society of Wetland Scientists: Research Brief*: http://1–5. www.sws.org/.../stein_2009_0010.pdf.

Stewart, David W., Prem N. Shamdasani, and Dennis W. Rook. 2007. *Focus groups: Theory and practice*. Vol. 20, Applied Social Research Methods series. 2nd ed. Thousand Oaks, CA: Sage.

Stone, Linda, and J. G. Campbell. 1984. The use and misuse of surveys in international development: An experiment from Nepal. *Human Organization* 43 (1): 27–37.

Taplin, Dana H., Suzanne Scheld, and Setha M. Low. 2002. Rapid ethnographic assessment in urban parks: A case study of Independence National Historical Park. *Human Organization* 61 (1): 80–93.

Thieme, T. 2010. Youth, waste and work in Mathare: Whose business and whose politics? *Environment and Urbanization* 22 (2): 333–52.

Thompson, L. 1970. Exploring American Indian communities in depth. In *Women in the field: Anthropological experiences*, edited by P. Golde, 45–64. Chicago: Aldine.

Toness, A. S. 2001. The potential of participatory rural appraisal (PRA) approaches and methods for agricultural extension and development in the 21st century. *Journal of International Agricultural and Extension Education* 8 (1): 25–35.

Tracy, Sarah J. 2013. Qualitative research methods: Collecting evidence, crafting analysis, communicating impact. Chichester, UK: Wiley-Blackwell.

Traugott, Michael W. 2005. The accuracy of the national preelection polls in the 2004 presidential election. *Public Opinion Quarterly* 69 (5): 642–54.

Trotter, Robert, Richard H. Needle, Eric Goosby, Christopher Bates, and Merrill Singer. 2001. A methodological model for rapid assessment, response, and evaluation: The RARE program in public health. *Field Methods* 13 (2): 137–60.

Trotter, Robert, and Merrill Singer. 2005. Rapid assessment strategies for public health: Promise and problems. In *Community intervention and AIDS*, edited by Edison Trickett and Willo Pequeqnat, 130–52. Oxford: Oxford University Press.

Umans, L. 1997. The rapid appraisal of a knowledge system: The health system of Guarani Indians in Bolivia. *Indigenous Knowledge and Development Monitor* 5 (3): 11–14.

Utarini, Adi, Anna Winkvist, and Gretel H. Pelto. 2001. Appraising studies in health using rapid assessment procedures (RAP): Eleven critical criteria. *Human Organization* 60 (4): 390–400.

Van Maanen, J., P. K. Manning, and M. L. Miller. 1998. Series editors' introduction. In *Doing team ethnography: Warnings and advice*, edited by K. Erickson and D. Stull, vi. Thousand Oaks, CA: Sage.

———. 1996. Series editors' introduction. In *Insider/outsider team research*, edited by J. M. Bartunek and M. R. Louis, v–vi. Thousand Oaks, CA: Sage.

Van Willigen, John, and Timothy L. Finan. 1991. *Soundings: Rapid and reliable research methods for practicing anthropologists*. NAPA bulletin. Vol. 10. Washington, DC: American Anthropological Association.

Wadsworth, Yoland. 2011. *Do it yourself social research: The bestselling practical guide to doing social research projects*. 3rd ed. Walnut Creek, CA: Left Coast Press.

Walsh, F., J. Abi-Nader, and M. I. Poutiatine. 2005. What do adult learners experience in a teacher certification program? *Journal of Adult Education* 34 (1): 6–21.

Walter, J. A., I. Coulter, L. Hilton, A. B. Adler, P. D. Bliese, and R. A. Nicholas. 2010. Program evaluation of total force fitness programs in the military. *Military Medicine* 175 (Supplement 1): 103–9.

Werner, Oswald, and Donald T. Campbell. 1970. Translating, working through interpreters, and the problem of decentering. In *A handbook of method in cultural anthropology*, edited by R. Naroll and R. Cohen. New York: Columbia University Press.

Westphal, Lynne M., and Jennifer L. Hirsch. 2010. Engaging Chicago residents in climate change action: Results from rapid ethnographic inquiry. *Cities and the Environment* 3 (1): 1–16. http://escholarship.bc.edu/cate/vol3/iss1/13.

Wheeler, Edyth J., and Walter D. Mallory. 1996. Team teaching in educational research: One solution to the problem of teaching qualitative research. Paper presented at the Annual Meeting of the American Educational Research Association. New York. EDRS document.

Willis, Jerry, with Muktha Jost and Rema Nilakanta. 2007. *Foundations of qualitative research: Interpretive and critical approaches*. Thousand Oaks, CA: Sage.

Wilson, Kathleen Karah, and George E. B. Morren. 1990. *Systems approaches for improvement in agriculture and resource management*. New York: Collier Macmillan.

Wolcott, Harry F. 2008. *Ethnography: A way of seeing*. 2nd ed. Lanham, MD: AltaMira.

———. 2005. *The art of fieldwork*. 2nd ed. Walnut Creek, CA: AltaMira.

———. 1999. *Ethnography: A way of seeing*. Walnut Creek, CA: AltaMira.

———. 1995. *The art of fieldwork*. Walnut Creek: AltaMira.

———. 1994. *Transforming qualitative data: Description, analysis, and interpretation*. Thousand Oaks, CA: Sage.

Woodside, Arch G. 2010. *Case study research theory, methods, practice*. Bingley, UK: Emerald.

REFERENCES

World Bank. 2001. *Stakeholder analysis*. http://www1.worldbank.org/publicsector/anticorrupt/PoliticalEconomy/stakeholderanalysis.htm.

Yin, Robert K. 2014. *Case study research: Design and methods*. 5th ed. Los Angeles, CA: Sage.

———. 2009. *Case study research: Design and methods*. 4th ed. Los Angeles, CA: Sage.

———. 2003. *Case study research: Design and methods*. 3rd ed. Thousand Oaks, CA: Sage.

———. 1994. *Case study research: Design and methods*. 2nd ed. Newbury Park, CA: Sage.

AUTHOR INDEX

SUBJECT INDEX

Note: Page numbers in italics indicate figures.

ABOUT THE AUTHOR

James Beebe is an applied qualitative social scientist, educator, and social activist committed to using the tools of the social sciences to promote justice. He is particularly concerned with developing tools such as Rapid Qualitative Inquiry that can be used for collaboration on identifying and addressing issues. In 2013 he founded and now directs the Center for Rapid Qualitative Inquiry, Global Networks in Portland, Oregon. In 2014 he was appointed a visiting researcher in the Anthropology Department of Portland State University.

He brings to his work an international reputation for helping define and popularize rapid qualitative research such as Rapid Assessment Process (RAP). His career has combined being a scholar with experience as a practitioner, each role influencing the other. More than twenty years of his experience has been outside the United States. James's career has included more than twenty years involvement with international development as a Peace Corps Volunteer (PCV), researcher, consultant, and Foreign Service Officer with the United States Agency for International Development and twenty years in academia including service as a professor of leadership studies at Gonzaga University, visiting special instructor of public policy at Kabul University (Afghanistan), professor of anthropology at Oregon State University, and associate professor at the Monterey Institute of International Studies. While serving as a PCV James taught at Mountain Province Community College and Columban College in the Philippines. He has delivered invited lectures at numerous universities, including Cornell, Harvard, UCLA, Washington State University–Vancouver, University of

Portland, Portland State University, Oregon Health and Science University, Colorado State University, University of Florida, University of the Philippines, Ateneo de Manila University, University of Botswana, and University of Pretoria.

While a PCV, James started graduate work in anthropology at the University of the Philippines. He completed a PhD in international development education, an MA in anthropology, and an MA in food research (international agricultural development) at Stanford University. For his doctoral dissertation, he did a year of fieldwork on farmers in a village in the Philippines. He also completed an M.Div. at Meadville Lombard, an affiliate of the University of Chicago.

James had long-term assignments with USAID in Sudan, Philippines, Liberia, and South Africa and short-term assignments in eleven other countries. In 2003 he was a Fulbright senior specialist in the Philippines. He has served on the board of the National Association for the Practice of Anthropology and is a fellow in the Society for Applied Anthropology.

As of 2014 he was involved in teaching, consulting, writing about his Peace Corps experience, and research focused on the application of anthropology to the analysis of U.S. foreign assistance policy in South Africa.